U0149997

# "一带一路"

## 瓜达尔地区热带干旱经济植物志

Non-wood Forest of The Belt and Road Initiative
Tropical Arid Area in Gwadar

王森 等／编著

中国林业出版社 China Forestry Publishing House

图书在版编目（ＣＩＰ）数据

"一带一路"瓜达尔地区热带干旱经济植物志 / 王

森等著. -- 北京：中国林业出版社，2023.11

ISBN 978-7-5219-1880-9

Ⅰ.①一… Ⅱ.①王… Ⅲ.①热带－干旱区－经济植

物－植物志－巴基斯坦 Ⅳ.①Q948.513.53

中国版本图书馆CIP数据核字(2022)第175119号


策划编辑：贾麦娥
责任编辑：贾麦娥
装帧设计：刘临川　张丽


出版发行：中国林业出版社
　　　　　（100009，北京市西城区刘海胡同7号，电话83143562）
电子邮箱：cfphzbs@163.com
网址：www.forestry.gov.cn/lycb.html
印刷：北京博海升彩色印刷有限公司
版次：2023年11月第1版
印次：2023年11月第1次
开本：889mm×1194mm　1/32
印张：12.75
字数：460千字
定价：168.00元

# 瓜达尔地区热带干旱经济植物种质资源调查收集活动发起人

王森（1972—），男，教授，博导，中南林业科技大学林学院副院长。2018年应中国海外港口控股有限公司之邀到巴基斯坦瓜达尔港进行林业生态环境建设咨询时，鉴于瓜达尔独特的气候条件、特殊的植被和当地产业结构，提出构建热带干旱经济林研究平台。并立即与育林控股集团有限公司付国赞教授、张赛阳展开种质资源调查收集工作。

张保中（1963—），男，客座教授，中国海外港口控股有限公司董事长、总经理。2014年带领公司从新加坡港务局接收瓜达尔的运营权，当看到满目荒漠的瓜达尔港时，表示要将热带干旱荒漠气候下的瓜达尔建成花园式港口，为港口职工和来往舰船船员提供优美的工作、生活、休养环境的建设蓝图，显示中国人民热爱绿色和平的决心。2021年1月受聘为中南林业科技大学客座教授。

胡华敏（1970—），女，客座教授，育林控股集团有限公司董事长。2016年应中国海外港口控股有限公司之邀到巴基斯坦瓜达尔港进行花园港口建设工作，2018年10月带领公司研究院付国赞教授/院长、张赛阳硕士/副院长，与中南林业科技大学、中国海外港口控股有限公司签署合作构建热带干旱经济林研究机构路线图。2019年8月受聘为中南林业科技大学林学院客座教授。

# 内容简介

本书对巴基斯坦瓜达尔地区主要热带干旱经济植物种质资源进行调查，分别从植物形态、生态分布、生长习性、用途价值等4方面进行系统整理。

本书紧密结合"一带一路"中巴经济走廊瓜达尔地区生态建设与产业发展的实际需要，以当地特色地带性植物为主、引入植物为辅，分乔木、灌木、草本、竹藤类4个章节进行描述。这项工作一方面弥补了我国对该类气候条件下经济植物研究不足的情况；另一方面为巴基斯坦瓜达尔地区生态建设和林业产业建设奠定了基础。

全书基础性、实用性并重。既可作为相关专业大学生、研究生的学习参考，亦可供从事生物学、林学、经济林学等科研人员参考。

## Introduction

The flora is aimed at investigating germplasm resources of non-wood plants that mainly grow in tropical arid areas in Gwadar, Pakistan. Also, it offers a well-organized introduction from four aspects in an organized way: plant morphology, ecological distribution, growth habit, and applications.

This flora closely combines the actual needs of ecological construction and industrial development in the Gwadar region that is related to the China-Pakistan Economic Corridor of "The Belt and Road" initiative. Characterized by native plants as well as supplemented by non-native plants, the flora introduces plants in detail and divides them into 4 chapters: trees, shrubs, herbs and bamboo & rattan. On the one hand, such work sufficiently compensates for the lack of research on non-wood plants under such climatic conditions in China due to the lack of tropical arid climate; on the other hand, it lays a foundation for the ecological construction and development of forestry industry in Gwadar, Pakistan.

Based on its elementary knowledge and practical applications, the flora can not only be used as a reference for college students and postgraduates in related majors, but also for scientific researchers engaged in biology, forestry, non-wood forests, and plant germplasm resources.

# 编辑委员会

主　　任　陈幸良　张保中

副 主 任　谭晓风　高　岩　胡华敏　李建安　郭文霞　冯彩云
　　　　　钟秋平　谭新建　王　森　Khalid Amin Sheikh（巴）
　　　　　Zhiming Liu(美)


主　　编　王　森　胡华敏　张赛阳

副 主 编　邵凤侠　巫　涛　付国赞　王珊珊　戚智勇

编　　委　梁文斌　李家湘　何功秀　李　泽　彭继庆　吴　毅
　　　　　陈　娟　葛晓宁　李　凯　郭　川　范艳霞　王利民
　　　　　闻　丽　赵天娇　何　钰　颜艳龙　吴大凯　王成玉
　　　　　谭玉珊　何铁定　王晨昊　付翰钰　吴博霖　王锦昊
　　　　　张惟楚　Yousuf Adnan（巴）　Irfan Ahmad（巴）

摄　　影　王　森　张赛阳　付国赞

翻　　译　王晨昊

# Editorial Board

**Director**

Chen Xingliang　Zhang Baozhong

**Deputy Director**

Tan Xiaofeng　Gao Yan　Hu Huamin　Li Jianan　Guo Wenxia　Feng Caiyun
Zhong Qiuping　Tan Xinjian　Wang Sen
Khalid Amin Sheikh (Pakistan)
Zhiming Liu (USA)

**Chief Editor**

Wang Sen　Hu Huamin　Zhang Saiyang

**Associate Editor**

Shao Fengxia　Wu Tao　Fu Guozan　Wang Shanshan　Qi Zhiyong

**Editorial Board Member**

Liang Wenbin　Li Jiaxiang　He Gongxiu　Li Ze　Peng Jiqing　Wu Yi
Chen Juan　Ge Xiaoning　Li Kai　Guo Chuan　Fan Yanxia　Wang Limin
Wen Li　Zhao Tianjiao　He Yu　Yan Yanlong　Wu Dakai　Wang Chengyu
Tan Yushan　He Tieding　Wang Chenhao　Fu Hanyu　Wu Bolin　Wang Jinhao
Zhang Weichu　Yousuf Adnan (Pakistan)　Irfan Ahmad (Pakistan)

**Photographers**

Wang Sen　Zhang Saiyang　Fu Guozan

**English Translators**

Wang Chenhao

# 序言一

## ——孟春之月，盛德在木

　　中南林业科技大学经济林科技创新团队的专家、教授、博士，多次亲临干旱酷热、几近荒漠的瓜达尔地区进行科学考察和现场试验，在瓜达尔成功种植了数百种耐高温、抗盐碱、御风沙、抵干旱的热带经济树种，并在此基础上编纂了《"一带一路"瓜达尔地区热带干旱经济植物志》，指导瓜达尔地区今后的绿化工作。付梓之时，王森教授和张赛阳博士希望我写篇感想，作为这本重要文献的序言。初始接到这个任务，确有诚惶诚恐之感觉，我对植物学研究一无所知，唯恐贻笑大方。然恭敬不如从命，更何况作为该著作的第一受益人，推广此书并借机表达对专家们的感激之情也是我应尽之责任。

## 一、瓜达尔：其山，其水，其人

　　瓜达尔位于巴基斯坦俾路支省西南端，面积1193km²，人口约21万。瓜达尔因港得名，由于其位于具有重要战略意义的波斯湾的咽喉附近，距离全球石油供应的主要通道霍尔木兹海峡大约400km，更由于中国的参与，瓜达尔声名显赫，引起人们的无限遐想。

　　悠悠几千年，瓜达尔一直默默无闻。巴基斯坦曾经是大英帝国在南亚地区的基地，但侵略者们似乎更加青睐巴基斯坦其他柔美茂盛的牧场，而对其西南部以茫茫尘沙和参差不齐的土山丘陵见长的瓜达尔地区鲜有兴趣。因此，阿拉伯海对岸的阿曼苏丹自1792年起才能够偏偶一方，占据瓜达尔200多年。直到1958年，巴基斯坦以300万英镑的价格自阿曼购买了瓜达尔地区，并入自己的版图。

　　2015年年底，公司委派我负责瓜达尔港口的运营开发工作。之前，对瓜达尔，我一无所知。当我从繁华都市北京，跨过蓝天碧海的印度洋降落在瓜达尔机场时，却被眼前的景象彻底震撼了：烫脸的热带季风，遮天蔽日的黄沙、了无生机的漫漫沙漠……曾经幻想憧憬的海边小城的浪漫柔情被这猝不及防的长河落日、大漠孤烟景象破坏得支离破碎。恍惚间，我怀疑飞机是不是把自己带到了火星！

瓜达尔属热带沙漠气候，年平均气温22~29℃，全年降雨量不到200mm。瓜达尔的当地语意思是"风之门"，风沙活动为其主要气候现象。蒸发量远远大于降水量，大风吹蚀，土壤盐渍化破坏了脆弱的植被生态系统，并形成恶性循环。如果不尽快进行人工干预，瓜达尔将很快变成一片沙漠。

在这片贫瘠的土地上生活着勤劳、勇敢、热爱和平的俾路支人。他们世世代代以打鱼为生，无法得到巴基斯坦联邦政府的财政支持。因此，这里的基础设施条件极其落后，老百姓们仍然过着缺医少药、无水无电的中世纪生活。可以说，瓜达尔是一个被现代世界遗忘的角落。

## 二、瓜达尔港：过去，现在，未来

早在20世纪70年代，巴基斯坦政府就有了开发瓜达尔港的设想。美国、俄罗斯都曾经试图染指瓜达尔港开发，但由于各种原因，无果而终。1999年穆沙拉夫总统执政后，请求中国政府援建瓜达尔深水港项目。穆沙拉夫于2000年7月对中国进行了友好访问，两国在一系列问题上达成了广泛的共识。中国于2001年作出了援建决定。当年8月，中巴两国政府在北京签署了瓜达尔港项目一期工程融资协议。2002年，中国港湾集团承担瓜达尔港工程建设，2005年完成工程建设。2007年，新加坡港务局获得该港口的运营开发权，但由于各种原因，港口运营没有获得预想效果，2013年，中国海外港口控股有限公司接管了瓜达尔港运营开发权。

2015年4月21日，习近平主席访问巴基斯坦，提出中巴两国要"以中巴经济走廊建设为中心，以瓜达尔港、能源、基础设施建设、产业合作为重点，形成'1+4'合作布局。让发展成果惠及巴基斯坦全体人民，进而惠及本地区各国人民"的倡议，瓜达尔获得了前所未有的发展机遇。

## 三、孟春之月，盛德在木

中国企业走出去30多年，在世界各地建造了无数个丰碑式的港口、公路、机场、电站、高楼大厦。但像瓜达尔这样不仅仅负责整个港口和自由区的建设，还负责运营管理，甚至主导整个地区今后若干年的经济和社会发展的综合性项目，还是第一次。如何在发展瓜达尔经济的同时，解决好民生和环境问题是摆在我们面前的一道综合考试必答题。

工业化是瓜达尔未来繁荣和发展的必然道路。但面对该地区恶劣的自然环境，脆弱的生态体系，我们必须在重视经济发展的同时，高度关注环境保护，使工业文明和生态文明融合发展；我们必须避免使瓜达尔在未来的工业化道路上陷入西方、包括中国部分地区"先污染、后治理"的魔咒；我们必须统筹协调，标本兼治，探索出一条代价小、效益好、排放少、可持续发展能力强的瓜达尔工业化发展道路。只有这样，我们才会一笔一画写好瓜达尔综合开发这个答卷；才会不负时代之托，把瓜达尔打造成为"一带一路"上的典范力作。

中国公司"发展经济，环保优先"的理念得到巴基斯坦政府的高度认可和支持以及社会各界的广泛支持。2019年3月底第二届北京"一带一路"高峰论坛会议期间，中国海外港口控股有限公司与巴基斯坦气候变化部签署了清洁绿色瓜达尔发展合作协议，涉及低碳环保、清洁能源、水资源开发和利用、生物多样性等多个领域，并提出在未来5年内开展大规模植树造林活动，种植不少于1000万株各种苗木，改善瓜达尔地区整体环境的具体要求。为配合中国港控执行好合作协议，中南林业科技大学多次派遣科研小组，对瓜达尔及其周边地区的土壤、植被进行深入科研考察，对中国港控的植树造林活动给予科学指导。

中巴经济走廊实施五年来，中国公司在瓜达尔筑港修路，兴建学校、医院、机场、电站，昔日凋敝破落、人迹罕至的瓜达尔如今已经成为世界各地投资商趋之若鹜的投资热土。中国公司带来的远不止工商业的逐步繁荣，更是生存环境的大幅度改变。短短五年间，瓜达尔树绿了，花红了，蝴蝶翩翩而至，鸟儿林中放声歌唱。中巴建设者们硬是在人迹罕至的戈壁滩上建起了生机盎然的花园式港口、公园式工业园区。

这本《"一带一路"瓜达尔地区热带干旱经济植物志》是对瓜达尔地区植物研究以及这几年绿化工作的系统性总结。本书中大量照片和数据基于大规模野外考察和标本采集的第一手材料，包含了许多新信息、新内容，有很高的科学价值和现实指导作用，为瓜达尔今后的生态建设提供有力的帮助。

前人栽树，后人乘凉，千方百计为子孙后代谋福利，是我们中华民族繁衍生息，不断发展壮大的根本原因。我们企业"走出去"，为世界人民带来的应当不仅仅是商品和技术，更要传播我们优秀的中华文化。两年前，为我年逾百岁的父母贺岁，老父亲唯一的要求就是植树纪念，庇荫子孙。如今老人作古，但他所代表的那种为后代子孙谋福利的中华文化正如他亲手种植的大树一样，万古长青！

张保中

中国海外港口控股有限公司董事长

中南林业科技大学客座教授

2020年8月10日

# Preface 1

## A great virtue to plant in spring

The non-wood forest science and technology innovation team from Central South University of Forestry and Technology , including experts, professors and doctors, have visited Gwadar, which is dry, hot and nearly desert, for many times to conduct scientific investigations and field tests, and have successfully developed hundreds of non-wood tropical plant species with good resistance to high temperature, salt, dust-sand weather, and drought. To instruct the further afforestation in Gwadar, experts in the team compiled the flora of Gwadar based on their practical experiments and research in such a dry and deserted region. Before the publication, I received a request from Professor Sen Wang and Doctor Saiyang Zhang who hope me to write down my feelings about the flora as a significant preface. When I initially accepted the request with profound reverence and humility, I worried about my competence to offer a satisfying preface due to my lack of knowledge about phytology and being exposed to ridicule before specialists. However, it's better to accept deferentially than to decline courteously when showing my sincerity. Moreover, as the primary beneficiary of such great work, it is my responsibility to promote the flora and express my appreciation to experts.

### 1 Geography and culture of Gwadar

Gwadar locates on the southwestern coast of Baluchistan, Pakistan. It takes up an area of 1193 square kilometers with a population of 210000. Gwadar is famous for its port since it is close to the bottleneck of the Persian Gulf with great strategic importance and situates at a distance of approximately 400 kilometers from the Strait of Hormuz which plays a dominant role in supplying oil for the whole world. Besides, it is because of the investment from China that Gwadar becomes more distinguished than before, thus resulting in people's prospects for the promising market in Gwadar.

Gwadar appears to be unknown to people for thousands of years. Pakistan was once the headquarter in south Asia established by the British colonists. The invaders preferred to pay more attention to lush pastures in other regions of Pakistan rather than the Gwadar region in the southwestern part which contains vast deserts and rugged mountains, explaining why the Sultan of Oman could rule this area since 1792. It was not until Pakistan paid 3 million pounds for the Gwadar region in agreement with Oman in 1958 that the Gwadar was absorbed into this country again.

At the end of 2015, I was appointed to Gwadar Port to take charge of the operation and exploitation by the corporation. I honestly knew nothing about this place before arrival. Nevertheless, I was amazed by the scenery presented to me: face-burning tropical monsoon, sky-covered sandstorm, desolated desert, etc. upon landing at the airport of Gwadar after flying across the Indian Ocean from Beijing. My imagination about a romantic coastal town broke up suddenly when facing the grand sight which was similar to the description in a poem "Lonely smoke rises up to the sky among the vast desert, along with the sun sinking beneath the skyline." In a trance, I could not help doubting whether the plane flew me to Mars!

Gwadar has a hot desert climate with an average temperature ranging from 22-29°C characterized by low precipitation. Gwadar means "the gate of wind" in the local language since the wind blow and sand movement activity is the main weather phenomena. High evaporation, wind erosion, and soil salinization severely damage the ecology of vegetation in Gwadar, forming a viscous cycle. If manual intervention can't be involved as soon as possible, chances are high that the whole of Gwadar will rapidly transform into a desert.

There lives Baloch who is brave, hard-working, and peaceful in this barren land. They have been living mainly on fishing for centuries. Because of the sparse population, people who live here have no access to financial support from the government. Consequently, the condition is extremely backward and terrible in terms of infrastructure. People still live a medieval-like life, lacking medicine, water, and electricity. In another word, Gwadar is a place forgotten by the modern world.

## 2 Gwadar: Past, Present, and Future

Early in the 1970s, the government of Pakistan came up with ideas that aimed at Gwadar's development. Both the US and Russia attempted to be involved in the development of ports but ended up in failure on account of various reasons. When President Musharraf took over administrative control in 1999, he requested the Chinese government to aid with the project of the deep-water harbor in Gwadar. He made a friendly visit to China and reached a broad consensus on a series of problems for both sides. After that, China decided to assist Gwadar to construct and develop its port in 2001 while in August, governments mutually signed an agreement concerning the financing of the first-period Gwadar Port project. In 2002, CHEC (China Harbor Engineering Company Limited) took charge of the project and completed it in 2005. PSA (Port of Singapore Authority) owned the development authority on port operation in 2007. However, the operation did not reach the desired result, thus it was transferred to COHG (China Overseas Holding Group) later in 2013.

On 21st April 2015, President Xi visited Pakistan and proposed that the cooperation between China and Pakistan should be centered on the development of CPEC (China-Pakistan Economic Corridor) and emphasized the deep-water harbor, energy, infrastructure construction, and industry collaboration, forming a "1+4" cooperation patterns to share the benefit with Pakistani and further promote such advantage to all other countries. From then on, Gwadar seized an unprecedented opportunity to develop.

### 3 A great virtue to plant in the first month of spring

Enterprises in China have constructed numerous memorable harbors, highway networks, airports, power plants, and residential buildings abroad over the past 30 years. However, it is the first time China has to cope with cases like Gwadar. We need to be responsible for the construction of the whole port and free zone well, in addition to considering relevant management, even the comprehensive development of the economy and society in the whole region for subsequent years. How to resolve issues on both livelihood and environment when developing economy simultaneously remains a necessary question for us to answer.

Industrialization is a necessity for Gwadar to advance toward prosperity in the future. When facing a deteriorating environment and weak ecosystem, we must concentrate on economic development as well as environmental protection to ensure an integrative development for both industry and ecosystem; we must prevent future development of industrialization from being trapped into the dilemma "grow first, clean up later", which some western countries and partial areas in China experienced in the past; we must persist on laying out an overall plan and paying significant attention to both symptoms and causes, finding out a way of development characterized by low-cost, high-efficiency and strong sustainability. Only in such a way can we offer a satisfying result and not fail to live up to people's expectation that we shall rebuild Gwadar into a masterpiece on the Belt and Road initiative.

The concept "Protecting the environment is prior to the economic development." raised by us is highly recognized by local government and supported widely by the whole society. During the period of the second Belt and Road forum held in Beijing, in March 2019, COHG successfully signed a cooperative agreement with PMD (Pakistan Meteorological Department) concerning clean and environmental-friendly development in Gwadar, which involves various aspects such as environmental protection, clean energy, utilization of water sources and biodiversity. Besides, it was suggested that afforestation would proceed in the next five years on a large scale with the amount of no less than 10 million saplings, improving the environment around Gwadar. Companies from China have built supported facilities such as schools, hospitals, airports, and power plants since the CPEC has been carried out.

The destitute Gwadar that few people stepped into formerly has become a hotspot attracting investors' preference worldwide. What Chinese enterprises bring about is more than a growing industry and business, we also helped improve the living environment. In just five years, there exist green trees and bright-colored flowers covering this land along with dancing butterflies and pleasant singing birds. Laborious workers from China and Pakistan made a huge effort to build garden-like ports and industrial districts full of vitality in the gobi off the beaten track.

Non-wood Forest of the Belt and Road Initiative Tropical Arid Area in Gwadar summarizes the botanical research and afforestation in recent years systematically. Pictures and relevant data demonstrated in the flora which originates in the primary source obtained from wild investigations and specimen collections include plenty of new information and details with a high value of science and guidelines, definitely providing strong support for future development in Gwadar's ecosystem.

As an old saying goes in China: "The next generation reaps what the former one sowed." The reason why the Chinese nation can be sustainably thriving and growing large lays in pursuing the benefit for the next generation. What our enterprise focuses on when going global is not only aiding people around the world with products and technology but propagating our outstanding traditional culture. Two years ago, I came to celebrate my parents' 100th birthday. My father required me to plant a tree in memory of his birthday, blessing my children. Although he passed away, the tradition for the benefit of the next generation he symbolized will long live with prosperity!

**Zhang Baozhong**

Chairman, China Overseas Holding Group

Visiting Professor, Central South University of Forestry and Technology

August 10, 2022

# 序言二

## ——登上高山，眺望大海

　　全球有1/3以上的土地处于干旱和半干旱地区，主要分布在两半球副热带区，有世界38%的人口，其中93%的人口在发展中国家。我国干旱、半干旱地区约占全国土地面积的52%。干旱与半干旱地区基本上与易荒漠化地区吻合，风沙危害严重，生态环境脆弱，经济欠发达。干旱和半干旱地区面临缺水和土地荒漠化的威胁，解决好生态环境建设和林业生产发展之间的关系，是该区域一项长期而艰巨的任务。

　　巴基斯坦瓜达尔港位于俾路支省西南端，年平均气温22～29℃，全年降雨量约200mm，年蒸发量却在2000mm以上。自然环境恶劣，生态体系脆弱，属热带干旱荒漠气候。该区接近阿曼湾北方出口，面积1193km²。距离全球石油供应的主要通道霍尔木兹海峡大约400km，是"一带一路"中巴经济走廊的桥头堡。国家对"一带一路"高质量建设，发出了"共谋绿色生活，共建美丽家园；携手防治荒漠，共谋人类福祉"的号召。巴基斯坦瓜达尔港在自然条件下受降水的限制不能够形成森林，只有发展有群众参与的经济林，当地居民为了定期收获有经济价值的产品，主动愿意为其灌水和施肥，"绿色高质量发展和美丽家园建设的号召"才能够落到实处。干旱经济林工程技术在修复荒漠生态系统、重建荒漠特色产业、推动干旱与半干旱地区人口资源与环境可持续发展正向互动方面发挥了不可替代的重要作用。

　　正是在这样的背景下，2018年9月到2019年11月，我校王森教授两次受邀到巴基斯坦瓜达尔港进行生态建设咨询工作。他回来后告诉我，当时震撼于瓜达尔港的景观：荒漠、荒漠、还是荒漠，好不容易走到头则是大海，并提出中巴两国经济林科学家在这个地方建设一个热带干旱经济林工程技术研究中心，将会对瓜达尔以及同样气候类型下地区的经济林产业培育与发展具有很好的引领作用。我想如果中南林业科技大学结合自己的优势学科经济林和巴基斯坦卡拉奇大学、费萨拉巴德农业大学、印度河大学的植物、经济林等学科、专业，在瓜达尔建设一个经济林国际联合实验室或国际化经济林教育研究中心，将会有力推动中南林业科技大学经济林学科建设和国际化高层次人才的培养。

2021年1月，中国海外港口控股有限公司张保中董事长、育林控股有限公司胡华敏董事长一行来到中南林业科技大学。他们非常赞赏我校派往瓜达尔的王森团队所体现的中南林人诚朴坚毅的作风和求是求新的精神，协议构建"一带一路"经济林科技国际联合平台，培养具有国际化视野的"一带一路"国际化经济林科技人才，致力于瓜达尔港的绿色生态建设和瓜达尔地区经济林产业建设。张保中董事长、胡华敏董事长与我校座谈并深入交流后，愉快地签署了《中南林业科技大学、中国海外港口控股有限公司和育林控股有限公司共同建设"一带一路"热带干旱经济植物联合重点实验室》的战略合作协议。2021年5月，学校专门下达中南林发[2021]21号文正式批准成立"一带一路"热带干旱经济林工程技术研究中心。2021年10月26日，由中南林业科技大学申请，中国科学技术协会批准，中国林学会主办的"第一届中巴热带干旱经济林科技交流会议"在中国长沙和巴基斯坦瓜达尔同时连线召开。巴基斯坦林业和野生动物部常秘道斯汀在致辞中表示，该次会议是中巴两国科研机构和企业开展科技创新合作的具体行动，是瓜达尔港从引资建设到引智发展的重要标志，有利于推动瓜达尔港及"中巴经济走廊"生态和经济发展。中国驻卡拉奇总领事馆郭春水参赞认为，科技交流活动将为瓜达尔地区输入中巴两国科技力量，促进绿色产业培育，是"中巴经济走廊"和"一带一路"建设向绿色高质量发展迈进的标志。中国林学会陈幸良秘书长表示，中国缺失巴基斯坦瓜达尔地区的热带干旱气候，对其独特的耐干旱、耐贫瘠的经济植物种质进行挖掘与利用研究，填补了中国林业科学研究在这一领域的空白，中南林业科技大学积极响应国家"走出去"战略，聚焦瓜达尔地区产业发展的瓶颈问题，开阔了我国经济林学科的国际化视野。

2021年12月，学校派王森教授的5人经济林研究团队，带着必要的实验仪器、实验药品从学校出发前往巴基斯坦的瓜达尔港。他们5人前往达瓜达尔港的路上，在郑州受到了育林控股有限公司胡华敏董事长，河南省林业厅科技外事处尚忠海处长，河南农业大学林学院范国强院长的热情接待；在北京受到中国牧业集团党委书记郭亮和组织部部长沈静的关怀和帮助；在瓜达尔受到了中国海外港口控股有限公司董事长、总经理、我校客座教授张保中同志的精心指导和悉心照顾，我在此代表学校表示感谢！

王森团队在巴基斯坦瓜达尔期间聚焦开展耐干旱、耐贫瘠经济林种质资源收集与利用的基础研究，编著50余万字的《亚热带地区枣属植物研究》专著1部；与巴基斯坦卡拉奇大学、费萨拉巴德农业大学、印度河大学创刊主编*Journal of Arid Non-wood Forestry Research*第1卷第1期内部学术通讯；构建20亩的种质资源库、1500m²组织培养中心、2处特色经济林示范园；收集圣冠枣、无花果、大花田菁等26种热带干旱特色经济林种质资源，组织培养繁育兰花、香蕉等17万株组培苗，构建无花果、大花田菁经济林示范园30亩，为巴基斯坦瓜达尔及同类型气候条件地区示范兼容生态效益、经济效益和社会效益的绿色可持续的发展道路。2021年11月26日，由中南林业科技大学申请，湖南省科学技术协会批准，中国林学会主办的"第二届中巴热带干旱经济林科技交流会议"在中国和巴基斯坦连线召开。巴基斯坦驻华大使莫因哈克在致辞中高度赞扬了"一带一路"热带干旱经济林工程技术研究中心为促进两国在科技、农业和林业等领域合作而开展的工作。并表示巴基斯坦被宣布为习近平主席全球发展倡议下的优先国家，该倡议也关注包括科学和技术在内的各个领域的高质量发展。中国驻卡拉奇总领馆杨广元参赞表示，前不久中巴两国领导人在北京会面就未来中巴合作达成一系列重要共识，本次会议的召开恰逢其时，可以汇聚各方力量，共商未来中巴农业合作方向，期待本次会议能够诞生一批实实在在的成果，真真正正地造福中巴两国人民，中国林学会和中南林业科技大学在这个领域有着丰富的经验、先进的技术和许多成功的范例，可以为中巴相关合作做出重大贡献。中国林学会秘书长陈幸良在致辞中表示，第二届交流会议是在第一届中巴热带干旱经济林科技交流活动取得的平台建设成果的基础上，两国林业科技工作者进一步聚焦"干旱、荒漠"这一林业产业建设的难点重点，吹响经济林产业建设攻坚战的号角，中巴两国林业科技工作者的持续交流活动，必将惠及两国人民。

他们所做的事情受到了巴基斯坦驻中国大使莫因哈克，巴基斯坦林业与野生动物常秘道斯汀，中国驻巴基斯坦大使馆庞春雪公使，中国驻卡拉奇总领事馆李碧建总领事、郭春水参赞、杨广元参赞、中国林学会陈幸良秘书长、中国水利部沙棘中心卢顺光副主任的赞扬；受到了中国海外港口控股有限公司张保中董事长、育林控股有限公司胡华敏董事长的高度认可；被《人民日报》《央视新闻》《新华社》《中国绿色时报》专题报道，我也为中南林业科技大学能够参与到"一带一路"中巴经济走廊瓜达尔的生态建设和林业产业建设所取得的成绩感到骄傲。

王森团队将前期在瓜达尔地区干旱经济植物种质资源调查与收集的情况凝结呈现出来，我甚欣慰，有感而发，欣然为序。

王汉青

中南林业科技大学党委书记

2022年12月16日

# Preface 2

## Greater vision, greater future

Over one-third area of the land lies in arid and semi-arid regions around the world, which is mainly distributed in the subtropical regions of the two hemispheres. Such regions consist of 38% population in the world, of which 93% lives in developing countries. In China, arid and semi-arid areas account for about 52% of the country's land area, and such areas are prone to desertification, which results in severe blown sand hazards, a fragile ecological environment, and an underdeveloped economy. Threatened by desertification and water shortage, reaching a practical resolution between ecological development and industrial production in the long term is arduous.

Located at the southwestern tip of Baluchistan Province, Gwadar port possesses an annual rainfall of less than 300mm but an annual evaporation of more than 2000mm, with an average annual temperature ranging from 22-29°C. Therefore, the tropical arid climate contributes to a harsh environment and fragile ecosystem in this region. Close to the northern exit of the Gulf of Oman, the port with an area of 1193 square kilometers is about 400km away from the Strait of Hormuz, which functions as the main channel for global oil supply, playing a vital role in China-Pakistan Economic Corridor of the "Belt and Road" initiative as a bridgehead. In pursuing high-quality construction of the "Belt and Road" initiative, the Chinese government has proposed the idea: Joining together and controlling desertification to live green and better. Due to the limitation of low rainfall in Pakistan, it is a hardship for Gwadar Port to form forests naturally. Only when residents are positively involved in the development of non-wood forest, willing to irrigate and fertilize them to harvest products with high values regularly, can the policy advocating high-quality environmental development and building of a beautiful homeland be fully implemented. Hence, technologies for non-wood forest engineering projects under an arid climate have played an irreplaceable and significant role in rehabilitating ecosystems, rebuilding industries, and promoting the positive interaction between population resources and sustainable development of the environment in arid and semi-arid areas.

Against the backdrop of the development in Gwadar Port, Professor Sen Wang in Central South University of Forestry and Technology (CSUFT) was invited to Gwadar Port twice for consultation during the period from September 2018 to November 2019. After his return, he told me that he was shocked by the landscape of Gwadar Port: endless desert as well as the endless sea at the end of his sight. He proposed that building up a research center designed for tropical arid non-wood forest engineering in this place with the aid of both Chinese and Pakistani scientists would result in a leading effect on the cultivation and development of the non-wood forest industry in Gwadar and areas with the same climate. From my point of perspective, the disciplinary construction of non-wood forest in CSUFT and the cultivation of international high-level talents will be greatly promoted if our university combines its own advantages in the field of non-wood forest with those of University of Karachi, University of Agriculture Faisalabad, and Indus University to establish an international joint laboratory and educational center of the non-wood forest. Following this, I appointed Qin Lichun, the deputy secretary of the party committee in the CSUFT who paid attention to this work previously, as the direct contact.

In January 2021, Zhang Baozhong  and Hu Huamin , chairman of China Overseas Holding Group Co., Ltd and chairman of Yulin Holdings Co., Ltd. respectively, visited CSUFT. They highly appreciated the honesty, perseverance, truth-seeking, and innovation embodied by Professor Wang Sen's research team which was sent to the Gwadar Port. To better serve the development of ecosystems and the forestry industry in Gwadar, they agreed to establish an international joint platform for non-wood forest and develop talents with global outlooks. After an in-depth communication, Zhang Baozhong  and Hu Huamin  were pleased to sign the strategic co-operational agreement " 'Belt and Road' Joint Key Laboratory for Arid Non-wood Forest Held by CSUFT, China Overseas Holding Groups and Yulin Holdings ". In May 2021, CSUFT officially approve the establishment of the "Belt and Road" Tropical Arid Non-wood Forest Engineering and Technology Research Center. On October 26, 2021, with the application from CSUFT and the approval from China Association for Science and Technology, the "First China-Pakistan Tropical Arid Non-Wood Forestry Science and Technology Exchange Conference" sponsored by the Chinese Society of Forestry was held simultaneously in Changsha, China and Gwadar, Pakistan online. Dostain Jamali, permanent secretary of the Ministry of Forests and Wildlife of Pakistan, mentioned in his speech that the meeting was a solid basis for innovation cooperation between scientific research institutions and enterprises of China and Pakistan. In the meantime, it also symbolized the transition in the development of Gwadar Port from attracting investment to bringing in talents, which was conducive to promoting the ecological and economic development of Gwadar Port and the China-Pakistan Economic Corridor. Guo Chunshui , the counsellor of the Chinese consulate-general in Karachi believes that the scientific exchange will input the advanced technologies of both China and Pakistan to the Gwadar and promote the cultivation of green industries, which indicates a step towards green and high-quality development of the "China-Pakistan Economic Corridor" and the "Belt and Road" imitative. Meanwhile, Chen Xingliang, secretary-general of the Chinese Forestry Society, said that, for China Forestry Society, the research on drought-resistant and barren-resistant plants in Gwadar has filled the gap in this aspect due to the lack of tropical arid climate in China. CSUFT actively follows a "go global" strategy, focusing on the bottleneck of industrial development in the Gwadar region, and broadening the international vision.

In December 2021, a five-member research team led by Professor Wang Sen was sent to Gwadar Port instructed by CSUFT with necessary experimental instruments and reagents. On their way to Gwadar Port, they were warmly treated by Hu Huamin , chairman of Yulin Holding Co., Ltd., Shang Zhonghai , divisional director of the Science and Technology Foreign Affairs Department of Henan Forestry Administration, and Fan Guoqiang , dean of the Forestry College of Henan Agricultural University, in Zhengzhou. In Beijing, they were treated with solicitous help from Guo Liang, Secretary of the Party Committee, and Shen Jing, Director of the Organization Department in China Animal Husbandry Group. In Gwadar, they received meticulous guidance and care from Zhang Baozhong. Here, I would like to express my gratitude on behalf of CSUFT!

During their stay in Gwadar, Professor Wang Sen and his research team focused on the collection and utilization of drought-resistant and barren-resistant non-wood forest germplasm resources, compiling a 500000 -words monograph "Studies on Genus Ziziphus Mill. in Subtropical Region". Furthermore, the research team published an internal communication in "Journal of Arid Non-wood Forestry Research" Volume 1, Issue 1, which is the starting publication with the joint forces from Karachi University, University of Agriculture, Faisalabad, and Indus River University. The research team constructed a 20-acre germplasm resource bank, a 1500-square-meter tissue culture center, and 2 demonstration gardens exhibiting special non-wood plants; They also collected germplasm resources of 26 kinds of non-wood plants with arid characteristics such as Sacred Crown Jujube, *Ficus carica*, and *Sesbania grandiflora* (L.) Pers, cultivating 170000 plants like orchids and bananas via tissue culture and setting up a 30-mu non-wood forestry demonstration garden for *Ficus carica,* and *Sesbania grandiflora* (L.) Pers, which shows a green and sustainable development approach compatible with ecological, economic, and social benefits for Gwadar and areas with similar climate conditions in Pakistan. On November 26, 2021, with the application from CSUFT and approval from Hunan Association for Science and Technology. The "Second China-Pakistan Tropical Arid Non-wood Forest Science and Technology Exchange Conference" sponsored by the Chinese Forestry Society was successfully held online between China and Pakistan. In his speech, Pakistani Ambassador to China Moin Ul Haque highly appreciated the work carried out by the "Belt and Road" Tropical Arid Non-wood Forestry Engineering Technology Research Center which greatly promotes cooperation between the two countries in the fields of science, technology, agriculture, and forestry. He said Pakistan was declared as a priority country under President Xi's Global Development Initiative, which also focuses on high-quality development in various fields including science and technology. Counselor Yang Guangyuan , the Chinese Consulate General in Karachi said that in recent times, the leaders of China and Pakistan met in Beijing and reached a series of important consensus on China-Pakistan cooperation shortly. Hopefully, I expect that the conference which discussed the further plans for the development of agricultural cooperation between China and Pakistan at such an appropriate time will truly benefit the people of China and Pakistan with the joint forces from various aspects. Furthermore, the rich experience, advanced technology, and successful paradigms in forestry owned by the Chinese Society of Forestry and Central South University of Forestry and Technology will greatly contribute to the cooperation between China and Pakistan. Chen Xingliang, secretary-general of the Chinese Society of Forestry, said in his speech that based on the former conference, the second exchange conference focused on the difficulties in the development of forestry industries with the theme of "aridity and desert", sounding a clarion of call of an uphill battle for construction of non-wood forestry industries. The consistent communication about forestry technologies between Chinese and Pakistan scholars will benefit people in both countries.

The achievements made by Wang Sen's research team have been highly appreciated by Pakistani Ambassador to China Moin Ul Haque, Pakistani Permanent Secretary Dostain Jamali of the Ministry of Forests and Wildlife of Pakistan, Minister Pang Chunxue of the Chinese Embassy in Pakistan, Consul General Li Bijian , Counselor Guo Chunshui and Counselor Yang Guangyuan of the Chinese Consulate General in Karachi, Secretary-General of Chinese Forestry Society Chen Xingliang , and Lu Shunguang Deputy Director General, Management Center for Seabuckthorn Development, Ministry of Water Resource, CHINA. Also, the research team is highly recognized by Chairman Zhang Baozhong of China Overseas Holding Group Co., Ltd., and Chairman Hu Huamin of Yulin Holdings Co., Ltd as well as People's Daily, CCTV News, Xinhua News Agency, and China Green Times with special reports. I am so proud of the achievements made by CSUFT regarding participation in the ecological construction and forestry industry construction of Gwadar in the "Belt and Road" China-Pakistan Economic Corridor.

I am very pleased that Professor Wang Sen and his research team have exhibited preliminary findings about non-wood forest resources in Gwadar. Accompanied by such thoughts and feelings, I wrote this preface to show my joyfulness.

**Wang Hanqing**

Secretary of the Party Committee in CSUFT

December 16, 2022

# 前　言

　　热带干旱地区基本上与易荒漠化地区吻合，风沙危害严重，生态环境脆弱，经济欠发达，面临缺水和土地荒漠化的威胁。解决好生态环境建设和林业生产发展之间的关系，是该区域一项长期而艰巨的任务。经济林是一个开放系统，为了持续地获得经济林产品，劳动者就会不断向开放系统中输入水、肥和抚育所需劳动力，使其生态效益与经济效益的双重功能能得到体现，能从根本上解决干旱地区"绿水青山"和"经济增长"的矛盾，是保障该类型区域社会发展与生态文明构建的新动力。开展干旱地区经济林种质资源收集利用、新品种选育、节水抚育技术集成等方面的研究和推广，对于改善热带干旱地区环境，实现绿色经济可持续发展具有重要意义。

　　干旱经济林工程技术在修复荒漠生态系统、重建荒漠特色产业、推动干旱与半干旱地区人口资源与环境可持续发展正向互动方面发挥了不可替代的重要作用，以色列、巴基斯坦和我国新疆、甘肃、陕西、河南、云南等地的经济发展的瓶颈，均由经济林产业的发展而打破，但从世界全局来看，受水的限制，干旱半干旱地区的经济林产业发展仍较缓慢，面临环境和人口的双重压力，问题复杂，矛盾尖锐，具有许多不确定性。在干旱半干旱地区经济林产业发展的人才培养、科学研究、产业培育等方面的力量还比较分散。各干旱半干旱地区解决问题主要靠挖掘本区域的自然和经济潜力，缺少国内外、区域内外、行业内外的统筹协调。亟须我们将"一带一路"国家与地区的干旱经济林产业培育，提高到新的战略高度，对未来的干旱半干旱地区经济林产业发展进行系统性的顶层设计，切实解决面临的一些重大科技问题，加强系统集成研究，加快实施，实现绿色经济可持续发展。

　　瓜达尔位于巴基斯坦俾路支省西南端，接近阿曼湾北方出口，面积1193km$^2$。紧挨波斯湾，位于具有重要战略意义的波斯湾的咽喉附近，距离全球石油供应的主要通道霍尔木兹海峡大约400km。瓜达尔年平均气温22～29℃，全年降雨量200mm左右，属热带干旱沙漠气候。2018年9月，我受邀到巴基斯坦瓜达尔港进行生态建设咨询工作，当时震撼于瓜达尔港的景观：荒漠、荒漠、还是荒漠，好不容易走到头则是大海。我想如果我国和巴基斯坦的林业科研工作者，在这个地方建设一个热带干旱经济林工程技术研究中心，将会对瓜达尔以及同样气候类型地区的林业产业培育与发展具有很好的引领作用。张保中董事长和胡华敏董事长听后非常支持我的想法，随后立即协调巴基斯坦军方派出6名士兵和我、付国赞教授、张赛阳博士一起，展开前期种质资源调查工作。2018年10月，中南林业科技大学、中国海外港口控股有限公司、育林控股有限公司，在巴基斯坦瓜达尔共同发起成立中国-巴基斯坦热带干旱经济林工程技术研究中心。

2019年11月，我与胡华敏董事长和张赛阳博士在巴基斯坦奎达市向中国驻巴基斯坦大使馆王志华参赞汇报中国-巴基斯坦热带干旱经济林工程技术研究中心建设推进情况时，参赞高屋建瓴地建议研究中心名称拓展为"一带一路"热带干旱经济林工程技术研究中心。回国后我们与巴基斯坦卡拉奇大学、印度河大学、费萨拉巴德农业大学相继签订协议共同构建工程中心；2021年5月，中南林业科技大学正式下文批准成立"一带一路"热带干旱经济林工程技术研究中心。2022年1月12日，湖南省科技厅正式认定工程中心为湖南省"一带一路"热带干旱经济林国际联合研究中心。工程中心组建以来，聚焦开展耐干旱、耐贫瘠经济林种质资源收集与利用的基础研究，构建种质资源库、良种苗木繁育场、特色经济林示范园，可为巴基斯坦瓜达尔及同类型气候条件地区，示范兼容生态效益、经济效益和社会效益的绿色可持续的发展道路。4年间，我们持续开展瓜达尔地区经济植物的收集、调查工作，共整理撰写了40万字的《"一带一路"瓜达尔地区热带干旱经济植物志》，弥补我国经济林科研工作在热带干旱地区的空白，开阔中国经济林学科的国际化视野。

感谢中南林业科技大学林学院、林学学科的资助。感谢科技部"一带一路"创新人才交流外国专家局项目"中巴干旱经济林科技交流与研究平台建设"（项目编号：DI202029 001I）；湖南省重点研发计划"中国-巴基斯坦热带干旱经济林种质资源研究与利用"（项目编号：2022WK2021）；湖南省科协海智计划示范项目"第二届中巴热带干旱经济林科技国际学术交流会议"（项目编号：湘科协通[2022]19号2-2-1）；湖南省科协国际与区域科技交流项目"主办第三届中巴热带干旱经济林科技交流会议"（项目编号：2023SKX-KJ-06）；"吐鲁番葡萄、枣的抗旱耐热砧木与根际微生物菌株的筛选"（项目编号：2023WK4008）资助。

感谢中国驻巴基斯坦大使馆庞春雪公使、中国驻卡拉奇总领事馆李碧建总领事、中国海外港口控股有限公司张保中董事长对我们的关怀与鼓励。感谢巴基斯坦驻中国大使莫因哈克、卡拉奇大学校长、费萨拉巴德农业大学、印度河大学的评价与鼓励。

感谢国家林业和草原局科技司郝育军司长、刘庆新处长对我们的关心与鼓励。感谢中国林学会陈幸良秘书长、沈瑾兰副秘书长、郭文霞主任、冯彩云主任的关心与帮助。感谢湖南省科技厅国际合作与交流处黄进副处长；湖南省人社厅侯自芳处长、龙涛副处长的关心与帮助。感谢中国林业出版社贾麦娥编辑的热情悉心帮助。

2022年2月26日夜于巴基斯坦瓜达尔

# FOREWORD

Tropical arid regions have a preference for desertification, accompanied by severe blown-wind hazards, fragile ecological environment, underdeveloped economy, and threats of water shortage. It is a long-term and arduous task to achieve a balance between ecological environment construction and forestry production development. As an open system, non-wood forest needs laborers to continuously input water, fertilizer, and workforce into this system to sustain production. Hence, the efficacy of both economy and environmental protection could be reached, which radically solves the contradiction between "Lucid water and lush mountains" and "Economic growth". Also, the non-wood forest system acts as a new driving force to ensure social development and ecological civilization construction in arid regions. Carrying out the research in terms of promotion of the collection and utilization of non-wood forest germplasm resources under arid climate, the selection of new varieties, and technological integration of water conservation are of great significance in improving the environmental conditions in tropical arid regions and realizing the sustainable development of the environmental-friendly economy.

Non-wood forest engineering designed for arid climates has played an irreplaceable and important role in rehabilitating desert ecosystems, rebuilding industries with desert characteristics, and promoting the positive interaction between population resources and the sustainable development of the environment in arid and semi-arid areas. The bottlenecks in economic development in places such as Israel, Pakistan, and Xinjiang, Gansu, Shaanxi, Henan, Yunnan in China were all broken by the development of the non-wood forest industry. However, due to the limitation of water, the growth of the non-wood forest industries in arid and semi-arid areas is relatively slow, facing various difficulties such as constraints from both the environment and population, intractable problems, acute contradictions, and plenty of uncertainties in progress. Also, talent training cannot be well prepared for scientific research and industrial development in the field of arid non-wood forestry economy. The majority of arid regions are still depending on exploiting their own natural or economic potentials without overall coordination from home and abroad and relevant and non-relevant fields or industries. Therefore, we must place the development of the non-wood forestry economy in countries or regions involved in the "Belt and Road" initiative into priority. With a top-level design, it is practical for us to resolve possible conundrums during development and accelerate comprehensive research as well as implantation to realize the sustainable development of the green economy.

Located on the southwest coast of Balochistan, Pakistan, Gwadar Port is close to the northern exit of the Gulf of Oman, with an area of 1193 km². Approaching the Persian Gulf with great strategic significance, the port is about 400km from the Strait of Hormuz which is responsible for global oil supply. With a tropical desert climate, the average annual temperature in Gwadar ranges from 22-29°C and annual rainfall is less than 200mm. In September 2018, I was invited to Gwadar Port as a consultant. Upon arrival, I was shocked by what I saw in Gwadar: endless desert and endless sea at the end of my sight. It suddenly occurred to me that the foundation of a research center aimed at technologies of non-wood forest engineering initiatives would play a leading role in the development of non-wood forest under similar climate conditions. After hearing this, Chairman Zhang Baozhong and Chairman Hu Huamin supported my idea and immediately discussed with the Pakistani military to send 6 soldiers to protect, Professor Fu Guozan, Doctor Saiyang Zhang and I for the preliminary investigation in Gwadar. In October 2018, with the joint forces from Central South University of Forestry and Technology (CSUFT), China Overseas Holdings Group, and Yulin Holdings, the proposal of setting up The Belt and Road Engineering Research Center for Tropical Arid Non-wood Forest in Gwadar was put forward.

In November 2019, when Chairman Hu Huamin and Dr. Zhang Saiyang reported to Wang Zhihua, Counselor of the Chinese Embassy in Quetta, Pakistan in terms of the progress of the research center, he suggested that the name of the research center could be further expanded from a strategically advantageous perspective: " 'Belt and Road' Tropical Arid Non-wood Forest Engineering and Technology Research Center". After returning to China, cooperating with University of Karachi, University of Agriculture Faisalabad, and Indus University, we decided to set up an engineering center. In May 2021, CSUFT officially approved the establishment of the "Belt and Road" Tropical Arid Non-wood Forest Engineering and Technology Research Center. On January 12, 2022, the Department of Science and Technology of Hunan Province officially recognized the engineering center as the international joint research center of the "Belt and Road" Tropical Arid Non-wood forest initiatives. Since its establishment, the center has been focusing on basic research on the collection and utilization of drought-resistant and barren-resistant non-wood forest germplasm resources, building up germplasm resource banks, breeding gardens for high-quality seeds, and demonstration gardens with characteristic non-wood forests, which demonstrates a green and sustainable development path that is compatible with ecological, economic and social benefits. In the past 4 years, we kept collecting and investigating non-wood forests in Gwadar and managed to compile a 400000-word " Non-wood Forest of the Tropical Arid Area in Gwadar " to make up for the lack of studies on arid non-wood forest in China and broaden our scientific researchers horizons.

Here, I would like to express my thanks for the financial support from the College of Forestry, Central South University of Forestry and Technology. Also, I appreciate the funding from the "Belt and Road" innovative talents exchange project "Construction of China-Pakistan Arid Non-Wood Forestry Exchange and Technology Centre" (ID: Dl202029 0011) approved by the State Administration of Foreign Experts Affairs, Ministry of Science and Technology of the People's Republic of China, project "Research and Utilization of China-Pakistan Arid Non-wood Forest Germplasm Resources" (ID: 2022WK2021) of Hunan Key Research and Development Program and demonstration project "The Second China-Pakistan Tropical Arid Non-wood Forests Science and Technology Exchange Conference" raised by Home (Help Our Motherland through Elite Intellectual Resources from Overseas) program of Hunan Association for Science and Technology (ID:2022[Communication of Hunan Association for Science and Technology] No.19 2-2-1).

I would like to express my thanks for care and encouragement from the Minister Pang Chunxue of the Chinese Embassy in Pakistan, Consul General Li Bijian of Consulate General of China in Karachi and Chairman Zhang Baozhong of China Overseas Holding Group Co., Ltd. Thanks to Pakistani Ambassador to China Moin Ul Haque, President of University of Karachi, University of Agriculture Faisalabad, and Indus University for their evaluation and encouragement. We would also like to thank Director Hao Yujun and Director Liu Qingxin of the Science and Technology Department of National Forestry and Grassland Administration for their care and encouragement. Thanks to Secretary General Chen Xingliang , Deputy Secretary General Shen Jinlan , Director Guo Wenxia and Director Feng Caiyun of China Society of Forestry, for their care and help. Thanks to Huang Jin , Deputy Director of International Cooperation and Exchange Department of Science and Technology Department of Hunan Province; Director Hou Zifang and Deputy Director Long Tao of Human Resources and Social Security Department of Hunan Province for their care and help.

At the end of my foreword, I would like to express my gratitude to editor Jia Mai'e of China Forestry Publishing House for her sincere help.

**Wang Sen**

February 26,2022

Gwadar, Pakistan

# 目录
# CONTENTS

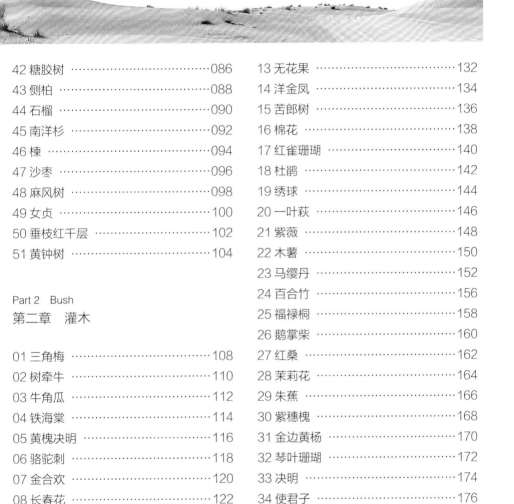

Part 2　Bush
## 第二章　灌木

第一章

# 乔木

PART 1　ARBOR

# 01 / 圣冠枣

*Ziziphus spina-christi*
鼠李科 Rhamnaceae 枣属 *Ziziphus*
别名：基督刺枣

## 植物形态

常绿乔木。树高4~8m。树皮灰褐色，纵裂。枝条呈"Z"字形。叶互生，卵圆形，深绿色，三出脉，钝齿，叶长2~6cm，宽1~4cm；针刺发达。花序呈聚伞花序；花黄绿色，直径4~6mm，花盘乳黄色；花期8~12月。果实扁圆形，直径5~20mm，红褐色；果期10月到翌年4月。种核圆形，直径2~5mm，灰褐色；种仁2枚。

## 生态分布

分布于以色列、巴基斯坦、阿富汗、印度、埃及等地。

## PLANT MORPHOLOGY

Evergreen tree. Plant height 4-8m. Bark grey-brown, longitudinally fissured. Branches zigzag-shaped. Leaves alternate, ovoid, dark green, three veins, crenate-like bicrenate, 2-6cm in length, 1-4cm in width; flourish thorn. Cymose inflorescences; flowers yellow-green, 4-6mm in diameter, disks whitish yellow; flowering period from August to December. Fruits oblate, 5-20mm in diameter, reddish brown; fruit period October to April. Stone globose, 2-5mm in diameter, taupe; kernels 2.

## ECOLOGICAL DISTRIBUTION

Distributed in Israel, Pakistan, Afghanistan, India, Egypt, etc.

## 生长习性

耐热，喜光。主根发达，再生能力强，对土壤适应性强，耐贫瘠、耐干旱、耐盐碱。

## 用途、价值

果实可食用。花可酿蜜。干果可治低血糖、高血压等症。果仁具有镇静作用。是热带干旱地区的特有经济植物。

## GROWTH HABIT

Heat-resistant, light-favored. Taproot flourished, strong regeneration, strong adaptability to soil, barren, drought and saline-alkali resistance.

## APPLICATION, VALUE

Fruit edible. Flowers can be made into honey; dried fruit can treat hypoglycemia, hypertension and other diseases; kernel effective for sedation; distinctive economic plant in arid regions.

## 02 / 枣

*Ziziphus jujuba*
鼠李科 Rhamnaceae　枣属 *Ziziphus*
别名：红枣、大枣、枣树

### 植物形态

落叶乔木。树高4~10m。树皮褐色或灰褐色，深裂。枝条分枣头、二次枝、枣股、枣吊等类型。叶互生；椭圆形、卵圆形至卵状披针形，深绿色，三出脉，具齿状锯齿，长2~10cm，宽1~8cm；针刺发达。花黄绿色，聚伞花序；萼片钝尖三角形；花瓣倒卵圆形；花盘乳白色、乳绿色、黄绿色；花期5~7月。果柱状、近圆球状、椭圆状、卵状或锥状，长15~60mm，直径14~40mm，成熟时红色，后变红紫色，果期8~10月。种核纺锤形或长椭圆形，直径2~8mm，黄褐色；常不具种仁，少见1~2枚。

### 生态分布

原产中国。主要分布于中国、韩国、日本、俄罗斯、保加利亚、澳大利亚、美国、加拿大、以色列、巴基斯坦。

### 生长习性

喜光，耐旱，耐涝性较强，开花期要求较高的空气湿度，否则不利于授粉坐果。对土壤适应性强，耐贫瘠、耐盐碱。

### 用途、价值

果实可鲜食，含有丰富的维生素。可制成红枣、蜜枣、熏枣、黑枣、酒枣

### PLANT MORPHOLOGY

Deciduous tree. Plant height 4-10m. Bark brown or grey-brown, fissure. Shoots including extension, secondary, mother bearing bearing, and so on. Leaves alternate, ovate to ovate-lanceolate or elliptic-oblong, dark green, three-veined, serrate, 2-10cm in length, 1-8cm in width, flourish thorn; Flowers yellow-green, inflorescence cymes; sepals ovate-triangular; petals obovate; flower disks milky, whitish green or yellowish green; flowering period from May to July. Fruits columnar, sub globose, ellipsoidal, ovoid or coniform, 15-60mm in length, 14-40mm in diameter, red at maturity, turning reddish purple, fruiting period from August to October. Stones spindly or oblong-elliptical, 2-8mm in diameter; yellowish brown, usually without kernel, narrowly 1-2.

### ECOLOGICAL DISTRIBUTION

Native to China. Mainly distributed in China, South Korea, Japan, Russia, Bulgaria, Australia, the United States, Canada, Israel and Pakistan.

### GROWTH HABIT

Thermophilic fruit tree; light-favored, strong drought and flooding resistance; higher humidity required during flowering period otherwise harmful to pollination and fruit setting. Strong adaptability to soil, Barren, salt-alkli resistance.

### APPLICATION, VALUE

Fruit edible, can be made into red jujube, candied jujube, smoked jujubes black jujubes, drunk jujube and dried jujube, succade, etc. Also available for jujube wine jujube vinegar, etc. Flowers made into honey; wood used for chariot making in acient

及干果、蜜饯。还可制成枣酒、枣醋等。花可酿蜜。枣木可制车、造船亦可雕刻成工艺品或制成乐器。枣可药用，有养胃、健脾、安神、滋补、强身等功效。枣树翠叶垂荫，果实累累，宜在庭园、路旁散植或成片栽植，具有防风固沙、降低风速、调节气温、防止和减轻干热风危害的作用，是干旱地区的主要经济植物。

time, shipbuilding, carving crafts or making musical instruments; applicable in medicine, effective in nourishing stomach, strengthening spleen, sedation, and building up body; shady trees with green leaves and numerous fruits, suitable for planting in gardens and roadsides or in patches, effective for wind prevention, sand stablization, wind speed reduction, temperature regulation, and prevention and mitigation of hot-dry wind; predominated economic tree in arid region.

# 03 / 椰枣

*Phoenix dactylifera*
棕榈科 Arecaceae　海枣属 *Phoenix*
别名：波斯枣、番枣、海棕、海枣

## 植物形态

常绿乔木。树高4~35m。茎灰褐色。叶直接着生于茎，呈稀疏头状树冠。叶长4~6m；叶柄长而纤细，多扁平；羽片线状披针形，长18~40cm，灰绿色。花白色，聚集为密集的圆锥花序；佛焰苞大而肥厚；花期3~4月。果长圆形或椭圆形，直径3~6cm；果肉肥厚；初生的果实呈青色，长大变为黄色，成熟后则呈红褐色或黄褐色；果期9~10月。种子扁平两端尖，腹面具纵沟；种仁1枚。

## PLANT MORPHOLOGY

Evergreen tree. Plant height 4-35m. Stem grey-brown. Leaves several on a stem, forming a sparse head-like crown, 4-6m in length; petiole long and slender, usually flat; pinna linear-lanceolate, 18-40cm in length, gray-green. Flowers white, clustered; inflorescence panicle; bract large and plump; flowering period from March-April. Fruit oblong or round, 3-6cm in diameter, fleshy; fruit cyan initially, turning yellow and then reddish-brown or yellowish-brown at maturity; fruiting period from September to October. Seeds flat, acuate at the apexes, longitudinal grooved from one side; kernel 1

## ECOLOGICAL DISTRIBUTION

Mainly distributed in Iran, Pakistan, Iraq, Saudi

## 生态分布

主要分布于伊朗、巴基斯坦、伊拉克、沙特阿拉伯、阿拉伯酋长联合国、以色列、埃及、利比亚等地。

## 生长习性

喜光，耐高温、耐霜冻，可在热带至亚热带气候下种植，对土壤要求不严，耐水淹、耐干旱、耐盐碱，但以土质肥沃、排水良好的有机壤土最佳。

## 用途、价值

果实含糖量高，营养丰富，富含多种维生素和天然糖分，可制成糖果、糖浆，也可用于酿酒与饮料制作；种子可榨油，炒焙后磨粉可作咖啡代用品；叶可做编织物和燃料；树干可以建造农舍和桥梁；果可入药，有消食、止咳化痰和治疗胃溃疡等功效。亦用于景观绿化，是热带干旱地区重要经济植物。

Arabia, The United Arab Emirates, Israel, Egypt, Libya, etc.

### GROWTH HABIT

Light-favored, heat-resistant and frost-resistant tree species. plantable in tropical to subtropical climates; not strict to the soil quality; flooding, drought and saline-alkali resistance, but organic, fertile loam with good drainage optimal.

### APPLICATION, VALUE

Fruit abundant in sugar, variable nutritions such as vitamin; applicable for producing candy and syrup, wine and beverge making also; the seeds squeezed to make oil; cooked seeds after grining used as alternative for coffee; leaves used as fabric and fuel; trunk used for farm and bridge building; fruit medicable, helping digestion, relieving cough and curing gastrohelcosis; ideal for landscape, important economic plant in tropical arid regions.

# 04 / 桤果木

## 植物形态

常绿乔木。树高3~15m。树皮灰褐色。树干笔直，嫩枝略带暗红色。叶互生；近革质，披针形，深绿色，羽状脉，长2~9cm，宽0.5~2.5cm。头状花序，无花瓣，银白色，被白粉，花序2~3簇生形成伞状花序。果实为多花聚合果，形状似纽扣，直径0.8~1.5cm。

## 生态分布

原产于索马里、吉布提和也门的沿海和河流地区。分布于巴基斯坦、尼泊尔、印度等地。

## PLANT MORPHOLOGY

Evergreen tree. Plant height 3-15m. Bark grey-brown. Trunk straight; shoots with minute dark red; leaves alternate, nearly leathery, lanceolate, dark green, pinnately-veined, 2-9cm in length, 0.5-2.5cm in width. Flower sliver, coverd with white powder, in heads, 2-3 clusters of inflorescence to form umbels; fruit aggregated, button-like shape, 8-15mm in diameter.

## ECOLOGICAL DISTRIBUTION

Native to coastal and river areas of Somalia, Djibouti and Yemen; distributed in Pakistan, Nepal, India, etc.

## GROWTH HABIT

Heat-resistant and saline-alkali resistant tree

## 生长习性

耐热耐盐碱树种。不耐寒，耐旱，喜温暖湿润的气候。在土层深厚、疏松、肥沃的土壤中生长良好。

## 用途、价值

可用于树篱、园林绿化、庭院栽培。是原生栖息地的再造林项目中的先锋树种。果实、幼树和芽可作饲料，木材可制炭。

species. Not cold-resistant but drought-resistant; favoring warm and humid climate; thriving in deep, loose and fertile soil.

## APPLICATION, VALUE

Used for hedges, landscaping and garden cultivation; pioneer tree species in the reforestation project of the native habitat; fruits, young trees and buds used as feed; wood made into charcoal.

# 05 / 辣木

*Moringa oleifera*
辣木科 Moringaceae　辣木属 *Moringa*
别名：鼓槌树

## 植物形态

常绿乔木。树高3~12m。枝具皮孔及叶痕，嫩枝被短柔毛。三回羽状复叶，羽片4~6对，羽片基部具线形或棍棒状稍弯腺体，常脱落，叶柄基部鞘状，小叶纸质，卵形或长圆形，长1~2cm，宽0.8~1.5cm，叶背苍白色。花白色，圆锥花序；花瓣匙形；花期全年。荚果，长20~60cm，直径1~3cm，表面具纵条纹，每荚含种子10~40粒，果期6~12月。种子近球形，褐色，直径约1cm，有3棱，每棱有膜质的翅。

## 生态分布

原产于印度。分布于中国、巴基斯

## PLANT MORPHOLOGY

Evergreen tree. Plant height 3-12m. Branches with lenticels and leaves with scars; new shoots puberulent. Leaves three-pinnate, pinnae in 4-6 pairs, clavate-curved or linear glands at base of pinnae, usually absent; petiole sheathing at base. leaflets papery, ovate or oblong, 10-20mm in length, 8-15mm in width, with pale back; flowers white, panicle; petals spoon-shaped; flowering period year round. Capsules 20-60cm in length,1-3cm in diameter, with longitudinal stripes on the surface; seeds 10-40; fruiting period from June to December. Seeds sub globose, brown, about 1cm in diameter, 3-angled, with membranous wings.

## ECOLOGICAL DISTRIBUTION

Native to India; distributed in China, Pakistan and pantropical habitants.

坦，广植于各热带地区。

## 生长习性

　　喜温喜光，适生于年平均气温21℃以上，极端最低气温3℃左右。可生长于半干旱地区；在土地肥沃、水分条件好的地方生长较快。能适应砂土和黏土等各种土壤，在微碱性土壤中能生长。

## 用途、价值

　　幼嫩果荚可食用。种子富含氨基酸、抗氧化剂、微量元素等，可提取润滑油；鲜叶可食用，老叶可制茶，是人类较古老的药食兼用的树种之一。材质较软，多用于薪柴，根部可用于雕刻。全年开花，用于观赏绿化、美化环境。

## GROWTH HABIT

Light-favored and thermophilus plant; suitable for temperature more than 21℃, but no less than 3℃; growing in semi-arid areas, thriving in places with fertile soil and good water conditions; adaptable to various soils including sand and clay, able to grow in slightly alkaline soil.

## APPLICATION,VALUE

Young capasules edible; seeds abundant in amino acids, antioxidants and microelement, from which lubricating oil extracted; young leaves edible, old leaves used for tea-making, one of the oldest tree species for both medical and edible application; texture soft, mostly used as firewood; root applicable in carving; flowering year round; cultivated as an ornamental tree.

# 06 / 桐棉

*Thespesia populnea*
锦葵科 Malvaceae  桐棉属 *Thespesia*
别名：恒春黄槿、杨叶肖槿

## 植物形态

常绿乔木。树高1~6m。树皮褐色。叶互生，纸质，卵状心形，全缘，掌状脉，先端长尾状，长7~18cm，宽4~10cm；叶柄长4~10cm，托叶线状披针形。花黄色，花冠钟形，基部具紫色块；花萼杯状，截形；花柱棒状；花期近全年。蒴果梨形，间有脉纹；被褐色纤毛，直径约4cm。种子三角状卵圆形，长约1cm。

## 生态分布

分布于中国、越南、印度、巴基斯坦、菲律宾及非洲热带地区。

## PLANT MORPHOLOGY

Evergreen tree. Plant height 1-6m. Bark brown. Leaves alternate, papery, ovate-cordate, entire-margined, venation palmate, acute to long-acuminate at the apex, 7-18cm in length, 4-10cm in width; petiole 4-10cm in length, stipules filiform-lanceolate. Flowers yellow, corolla campanulate, purple at the base; calyx cup-shaped, truncate; style columnar; flowering period nearly year round. Capsule pear-shaped, veined, brown tomentellous, about 4cm in diameter. Seeds triangular-ovid, about 1cm in length.

## ECOLOGICAL DISTRIBUTION

Distributed in China, Vietnam, India, Pakistan, Philippines and tropical regions of Africa.

## 生长习性

耐热树种，不耐寒，喜光；常生于海边和海岸向阳处。适生于排水力良好的中性至微碱性的壤土或砂质土。

## 用途、价值

树形高大，花形优美，具有绿化和观赏价值。全株具清热解毒、消肿止痛的功效；根可制成滋补品，树皮可用于治痢疾，果实分泌出的黄色黏液可治皮癣。

## GROWTH HABIT

Heat-resistant tree species, not resistant to cold, light-favored; common along sandy beaches; favored growing in neutral to slight alkaline soil or sand with good drainage.

## APPLICATION,VALUE

The tree is tall and the flower is beautiful, with greening and ornamental value; effective in fever-relieving, detoxifying, swelling-reducing and the root can be made into a tonic, the bark can be used to treat dysentery, and the yellow mucus secreted from the fruit can treat tinea.

# 07 / 凤凰木

## 植物形态

落叶乔木。树高5~20m。树皮灰褐色；小枝具皮孔常被短柔毛。二回偶数羽状复叶，具托叶，托叶羽状分裂或呈刚毛状；叶柄基部膨大呈垫状；小叶密集对生，被绢毛，基部偏斜。总状花序顶生或腋生，花鲜红或橙红色，花梗长3~10cm；花托盘状或陀螺状，花期6~7月。荚果带形，扁平，长30~60cm，宽3.5~5cm，红褐色，成熟时黑褐色，顶端有宿存花柱，果期8~10月。种子长圆状，长约15mm，宽约7mm，黄色，具褐斑，20~40枚。

## PLANT MORPHOLOGY

Deciduous tree. Plant height 5-20m. Bark gray-brown; branchlets with lenticels often pubescent. Two-fold even-numbered pinnate compound leaves, with stipules, pinnately divided or bristly; base of the petiole enlarged and cushion-shaped; leaflets densely opposed, sericeous, and base skewed. Racemes terminal or subaxillary flowers bright red or orange-red, peduncle 3-10cm in length; flowers tray-shaped or gyroscopic, flowering period from June to July. Pods ribbon-shaped, flat, 30-60cm in length, 3.5-5cm in width, reddish-brown, dark brown when mature, with persistent styles at the top, fruiting period from August to October. Seeds oblong, about 15mm in length and 7mm in width, yellow, with brown spots, 20-40 pieces.

## 生态分布

原产马达加斯加。分布于中国、巴基斯坦等地；世界热带地区常栽种。

## 生长习性

热带树种，喜高温湿润和阳光充足的环境，适生温度20~30℃，忌积水，耐干旱、耐瘠薄、不耐寒。适生于深厚肥沃、富含有机质的砂质土。

## 用途、价值

可作园林绿化观赏树种；根系有固氮根瘤菌；落叶量大，为地被植物提供良好的覆盖物，起到保湿保温、改善土壤有机质含量和结构、增加土壤肥力的作用。木材轻软，富有弹性和特殊木纹，可作家具和工艺原料；树脂可用于工业；豆荚可作打击乐器。

## ECOLOGICAL DISTRIBUTION

Native to Madagascar. Distributed in China, Pakistan, etc. Often planted in tropical regions of the world.

## GROWTH HABIT

Tropical tree species, favoring heat and humid and sunny environment, Suitable for temperature is 20-30°C, avoid standing water, barren and drought-resistance, not cold-resistant. Favored growing in deep fertile, sandy soil with organic matter.

## APPLICATION,VALUE

Used as a garden greening ornamental tree species; the root system has nitrogen-fixing rhizobia; the amount of fallen leaves is large, which provides a good cover for ground cover plants. Through ventilation and heat preservation, it can improve the content and structure of soil organic matter and increase soil fertility. The wood is light and soft, full of elasticity and special wood grain, which can be used as furniture and craft materials; resin can be used as industry; pods can be used as percussion instruments.

# 08 / 人心果

*Manilkara zapota*
山榄科 Sapotaceae　铁线子属 *Manilkara*
别名：吴凤柿、赤铁果、奇果

## 植物形态

　　常绿乔木。高2~20m。小枝褐色且具明显叶痕。叶互生，革质，密聚枝顶，长圆形或卵状椭圆形，全缘或微波状，网脉细密，长5~20cm，宽2.5~4cm。花冠白色，花冠裂片先端具不规则细齿，背面两侧具2枚花瓣状附属物。浆果，纺锤形、心状或球形，褐色，长4cm以上。种子扁状，花果期4~9月。

## 生态分布

　　原产美洲热带地区。分布于中国、印度、巴基斯坦等地。

## PLANT MORPHOLOGY

　　Evergreen tree. Plant height 2-20m. Branchlets brown, with obvious leaf scars. Leaves alternate, leathery, densely clustered at the top of the branches, oblong or oval-elliptic, whole or microwave-shaped, with fine mesh veins, 5-20cm in length and 2.5-4cm in width; Corolla white, with irregular fine teeth on the apex of the corolla lobes, and 2 petal-like appendages on both sides of the back. Berry, spindle-shaped, heart-shaped or spherical, brown, more than 4cm in length; seeds flat. Flowering and fruiting period from April to September.

## ECOLOGICAL DISTRIBUTION

　　Native to tropical America. Distributed in China, India, Pakistan, etc.

## 生长习性

喜高温多湿，不耐寒；适生温度22~30℃，适生于肥沃深厚的砂质或黏质土。

## 用途、价值

果可食，营养价值高，味甜可口；树干汁液为口香糖原料；种仁含油率高；树皮含植物碱。可用于绿化、观赏、盆栽。

## GROWTH HABIT

Favoring heat and humid, but not cold-resistant; suitable for temperature is 22-30°C, Favored growing in deep, fertile and sandy or clay soil.

## APPLICATION,VALUE

Fruit edible, with high nutritional value, sweet and delicious; the trunk juice is the raw material of chewing gum; the kernel has a high oil content; the bark contains plant alkaloids. Used for greening, ornamental and potted plants.

# 09 / 银合欢

*Leucaena leucocephala*
豆科 Leguminosae　银合欢属 *Leucaena*
别名：百合欢

## 植物形态

小乔木。树高2~6m，枝条具褐色皮孔，幼枝被短柔毛。二回羽状复叶，最下端羽片有1黑色腺体；长5~9cm，叶轴被柔毛。花白色，头状花序，花瓣倒披针形，长约0.5cm，背面被疏柔毛，直径2~3cm，花期4~7月。果为荚果，扁平带状，成熟时褐色，果期7~10月；种子卵圆形，大小约1cm，褐色，扁平，光亮，8~20枚。

## 生态分布

原产热带美洲。分布于中国、巴基斯坦、印度等地。

## PLANT MORPHOLOGY

Small evergreen tree. Plant height 2-6m, the branches have brown lenticels, young branches pubescent. Two pinnately compound leaves, with a black gland on the lowermost pinna; 5-9cm in length, with pilose shafts. Flowers white, with flower heads, petals oblanceolate, about 5mm in length, sparsely pilose on the back, 2-3cm in diameter, flowering period from April to July. Fruit pod, flat and ribbon-shaped, brown when mature, fruiting period from July to October; seeds ovoid, about 10mm in size, brown, flat, bright, 8-20 pieces.

## ECOLOGICAL DISTRIBUTION

Native to tropical America. Distributed in China, Pakistan, India, etc.

## 生长习性

喜温暖湿润、光照充足的环境，适生温度20~30℃。抗旱能力强，忌涝。适生于中性至微碱性土壤。

## 用途、价值

木材可制炭，是优良的薪炭柴树种。适合于荒山造林；是优良的多用途树种，绿化的优良树种，也可作观赏树种。叶可作饲料。种子可食，树胶作食品乳化剂或代替阿拉伯胶；亦可加工成杀菌剂。树皮可药用，用于治疗心悸、怔忡、骨折等症。

## GROWTH HABIT

Favoring warm, humid and sunny environment. Suitable for temperature is 20-30°C. Strong drought resistance, avoid waterlogging. Favored growing in neutral to slightly alkaline soil.

## APPLICATION,VALUE

Wood can be used to make charcoal, which is an excellent firewood species. It is suitable for afforestation in barren mountains. It is an excellent multi-purpose tree species. It can also be used as ornamental tree species. Leaves can be used as feed. The seeds are edible, and gum is used as food emulsifier or instead of arabic gum; it can also be processed into fungicide. Bark can be used for medicine, effective in treating palpitate, severe palpitation and fracture.

# 10 / 阔荚合欢

*Albizia lebbeck*
豆科 Leguminosae　合欢属 *Albizia*
别名：大叶合欢

## 植物形态

落叶乔木。树高6~10m；树皮灰白色，粗糙。嫩枝密被短柔毛，老枝无毛。二回羽状复叶，长6~20cm；小叶4~8对，长椭圆形，长2~4cm，宽1~2cm。花黄绿色，头状花序；花期5~9月。荚果带状，长10~30cm，宽2~4cm，扁平，褐色，常宿存于树上经久不落；果期10月至翌年5月。种子椭圆形，长约1cm，棕色，4~20枚。

## 生态分布

原产于热带非洲。分布于中国、埃及、索马里等地。

## PLANT MORPHOLOGY

Deciduous tree. Plant height 6-10m; bark gray white, rough. Young branches densely pubescent and t old ones glabrous. Secondary pinnate compound leaves, 6-20cm in length, Leaflets 4-8 pairs, long elliptic, 2-4cm in length, 1-2cm in width. Flower yellow green with head inflorescence; flowering period from May to September. Pods banded, 10-30cm in length and 2-4cm in width, flat and brown, often persist on the trees for a long time; fruiting period from October to May of the next year. Seeds elliptic, 10mm in length, brown, 4-20 seeds.

## ECOLOGICAL DISTRIBUTION

Native to tropical Africa. Distributed in China, Egypt, Somalia, etc.

## 生长习性

喜光树种，耐半阴，喜欢温暖湿润的生长环境，不耐寒；在肥沃和排水良好的土壤上生长良好。

## 用途、价值

木材优良，适用于家具、支柱、建筑等。叶子可作家畜的饲料；树形优美，具有一定的抗风、抗空气污染能力，是良好的园林绿化树和行道树。

## GROWTH HABIT

Intolerant trees, tolerant of half shade, favoring warm and humid environment, not cold-resistant; Favored growing in fertile soil with good drainage.

## APPLICATION,VALUE

Wood excellent, suitable for furniture, pillars, buildings, etc. The leaves can be used as fodder for livestock; the tree is beautiful in shape and has certain resistance to wind and air pollution. It is a good garden scenery, greening tree and street tree.

# 11 / 牧豆树

*Prosopis juliflora*

豆科 Leguminosae　牧豆树属 *Prosopis*

## 植物形态

常绿小乔木或灌木。树高2~15m。二回羽状复叶，羽片1~4对；小叶10~20对，长1~2cm，长圆形，基部偏斜，两面均被短柔毛或仅具缘毛，长1~2cm，托叶成对，刺状，常簇生于短枝上。花黄绿色，总状花序腋生，花萼杯状，顶端及边缘被毛，花瓣5。荚果线形，弯曲或劲直，无毛，长5~25cm，淡黄色，果期6月。种子长圆形，每个荚果含种子5~20枚。

## 生态分布

原产于热带美洲。分布于中国、巴

## PLANT MORPHOLOGY

Small evergreen tree or shrub. Plant height 2-15m. Two pinnately compound leaves, 1-4 pairs of pinnae; 10-20 pairs of lobules, 1-2cm in length, oblong, oblique at the base, both sides pubescent or only ciliate, 1-2cm in length, stipules paired, stab-shaped, often clustered on short branches. Flowers yellow-green, racemes axillary, calyx cup-shaped, top and edge hairy, petals 5. Pods linear, curved or straight, glabrous, 5-25cm in length, light yellow, fruiting period in June. Seeds oblong, each pod contains 5-20 seeds.

## ECOLOGICAL DISTRIBUTION

Native to tropical America. Distributed in China, Pakistan, India, etc.

基斯坦、印度等地。

## 生长习性

深根性树种，耐旱，能够生长于干旱、炎热地区；适应性强，对土壤和水分要求不高。

## 用途、价值

荚果可用作饲料，是重要的饲料植物。花可酿蜜。木材可用作铁路枕木、工艺雕像、乐器等，也是干旱地区重要的优良薪炭材。生命力强，防止土地退化和沙漠化，具有绿化和生态恢复作用。

## GROWTH HABIT

Deep-rooted tree species, drought-resistant, able to grow in arid and hot areas; strong adaptability, not strict to the soil and water quality.

## APPLICATION,VALUE

Pod can be used as feed, which is an important feed plant. Flowers make honey. Wood can be used as railway sleepers, craft statues, musical instruments, etc. it is also an important fine firewood in arid areas. It has strong vitality, prevents land degradation and desertification, and has the function of greening and ecological restoration.

# 12 / 鸡蛋花

*Plumeria rubra* 'Acutifolia'
夹竹桃科 Apocynaceae  鸡蛋花属 *Plumeria*
别名：缅栀子、蛋黄花

## 植物形态

落叶小乔木。高2~8m。枝条绿色肉质，具乳汁。叶厚纸质，长椭圆形，基部狭楔形，中脉在叶面凹入，在叶背略凸起，叶柄上面基部具腺体。聚伞花序顶生，花冠外面白色，花冠内面黄色；花期5~10月。蓇葖果双生，长圆形，形似羊角，种子长圆形，顶端具膜质的翅，翅长约2cm。

## 生态分布

原产墨西哥。分布于中国、巴基斯坦等地。

## PLANT MORPHOLOGY

Small deciduous tree. Plant height 2-8m. Branches green and fleshy, with sap. Leaves thick and papery, oblong, narrow and wedge-shaped at the base. Midvein concave on the leaf surface and slightly convex on the back of the leaf. The base of the petiole has glands. Cymes terminal, outer surface of the corolla white, inner surface of the corolla yellow; flowering period from May to October. Follicles twins, oblong, shaped like horns, seeds oblong, with membranous wings on the top, wings of about 2cm in length.

## ECOLOGICAL DISTRIBUTION

Native to Mexico. Distributed in China, Pakistan, etc.

## 生长习性

阳性树种，喜高温、湿润和阳光充足的环境，不耐寒，耐干旱，忌涝渍，抗逆性强。适生温度20~25℃。适生于深厚肥沃、富含有机质的酸性砂壤土。

## 用途、价值

广泛应用于公园、庭院、绿带、草坪等的绿化、美化。花晾干可作凉茶饮料和中药，具有清热解暑、润肺润喉，还有治疗咽喉疼痛等功效。亦可提取香精。木材质轻而软，可制作家具、乐器和餐具等。

## GROWTH HABIT

Intolerant trees, favoring heat, humid and sunny environment, not cold-resistant, drought-resistant, avoid waterlogging, strong resistance. Suitable for temperature is 20-25°C. Favored growing in deep, fertile, acidic sandy loam with organic matter.

## APPLICATION,VALUE

In landscape greening, used in parks, courtyards, green belts, lawns, etc. Dried flowers can be used as herbal tea drinks and traditional Chinese medicine. It has the effects of clearing heat, moistening the lung and throat, and treating sore throat. It can also extract essence. The wood material is light and soft, and can be used to make furniture, musical instruments and tableware.

# 13 / 大花田菁

*Sesbania grandiflora*
豆科 Leguminosae　田菁属 *Sesbania*
别名：红蝴蝶、木田菁

## 植物形态

小乔木。树高2~10m。羽状复叶，长10~40cm；小叶10~30对，长椭圆形，深绿色。总状花序，下垂，花冠粉红色至玫红色；旗瓣长圆状倒卵形至阔卵形，花萼绿色，长1~3cm。荚果线形，稍弯曲，下垂，长10~60cm，直径0.5~1.0cm。种子椭圆形，长约1cm，宽0.2~0.5cm，红褐色；花果期9月至翌年4月。

## 生态分布

原产马来西亚与印度尼西亚，分布于中国、越南、巴基斯坦等地。

## 生长习性

喜温暖、湿润的气候，不耐寒；适生于土层深厚、疏松、肥沃的土壤中。

## 用途、价值

花大而艳丽，用于园林绿化、造林、庭园观赏。花、叶可食用，木材造纸，制作软木塞，内皮提取优质纤维，也是一种药用植物，树皮可作收敛剂。

## PLANT MORPHOLOGY

Small tree. Plant height 2-10m. Pinnate compound leaves, 10-40cm long; leaflets 10-30 pairs, long elliptic, dark green. Racemes, pendulous, corolla pink to rose red; flag petals oblong obovate to broadly ovate, calyx green, 1-3cm long. The pods are linear, slightly curved, pendulous, 10-60cm long and 0.5-1.0cm in diameter; the seeds are elliptic, about 1cm long and 0.2-0.5cm wide, reddish brown; the flowering and fruiting period is from September to April of the following year.

## ECOLOGICAL DISTRIBUTION

Native to Malaysia and Indonesia. Distributed in China, Vietnam, Pakistan, etc.

## GROWTH HABIT

Favoring warm and humid climate, but not cold resistant; it is suitable for growing in deep, loose and fertile soil.

## APPLICATION,VALUE

The flowers are large and gorgeous, and are used for landscaping, afforestation and garden ornamental plants. Flowers and leaves are edible, wood is used to make paper, cork is made, and high-quality fiber is extracted from endothelium. It is also a medicinal plant, and bark can be used as astringent.

# 14 / 油橄榄

*Olea europaea*
木樨科 Oleaceae 木樨榄属 *Olea*
别名：洋橄榄、木樨榄

## 植物形态

常绿小乔木。树高2~10m。树皮灰色，散生皮孔。叶片革质，披针形或椭圆形；叶片被银灰色鳞片，全缘，边缘反卷；中脉两面凸起，叶柄密被银灰色鳞片，长1.5~6cm，宽0.5~1.5cm。花白色，圆锥花序，长2~4cm；花萼杯状；花冠长约0.5cm，深裂达基部；花梗长约10mm；花期4~5月。核果椭圆形，长1~4cm，成熟时蓝黑色，果期6~9月。

## 生态分布

原产小亚细亚。分布于地中海地区，以西班牙、意大利、希腊、葡萄

## PLANT MORPHOLOGY

Small evergreen tree. Plant height 2-10m. Barkgray with loose skin holes. Leaves leathery, lanceolate or elliptic; leaves covered with silver-gray scales, entire, edges recurved; midrib convex on both sides, petiole densely covered with silver-gray scales, 1.5-6cm in length and 0.5-1.5cm in width. Flower white, panicle, 2-4cm in length; calyx cup-shaped; corolla about 0.5cm in length, deeply divided to the base; pedicel about 0.1cm in length; flowering period from April-May. Drupe oval, 1-4cm length, blue-black at marturity, fruiting period from June to September.

## ECOLOGICAL DISTRIBUTION

Native to Asia Minor. Distributed in the Mediterranean region, with Spain, Italy, Greece,

牙、摩洛哥及阿尔巴尼亚为集中产地。

## 生长习性

喜温暖、湿润的环境，不耐旱；适生温度18~24℃。适生于含有石砾的中性或微碱性土壤。

## 用途、价值

果实可榨油，果实营养价值高，可食用。鲜果可作罐头和蜜饯，亦可制润滑剂、化妆品、肥皂等。木樨榄油具有食用和药用价值，能抑制肿瘤细胞生长，降低肿瘤发病率。可作绿篱、绿墙，用于园林绿化。

Portugal, Morocco and Albania as the concentrated production areas.

## GROWTH HABIT

Favoring warm and humid environment, but not drought tolerant; suitable for 18-24°C temperature. It is suitable for neutral or slightly alkaline soil containing gravel.

## APPLICATION,VALUE

Fruit can be pressed for oil, and it has high nutritional value and is edible. Fresh fruit can be canned and preserved, and can also be used as lubricant, cosmetics, soap and so on. Oleoresin has edible and medicinal value, which can inhibit the growth of tumor cells and reduce the incidence of tumor. Can be used as hedges and green walls for landscaping.

# 15 / 三色千年木

*Dracaena marginata* 'Tricolor'

天门冬科 Asparagaceae 龙血树属 *Dracaena*

别名：彩虹龙血树、三色缘龙血树

## 植物形态

常绿小乔木。树高1~3m。茎干灰褐色，呈圆柱形，挺拔直立。叶片簇生于茎干顶端，剑形，长10~60cm，宽1~2cm，绿色叶片上有乳白色、黄白色、红色的条纹。总状花序生于茎或枝顶端。浆果，种子1~3枚。

## 生态分布

原产马达加斯加。分布于中国、巴基斯坦等地。

## 生长习性

喜高温多湿和阳光充足的环境，耐

## PLANT MORPHOLOGY

Small evergreen tree. Plant height 1-3m, Stem dry gray brown, cylindrical, erect. Leaves clustered on the top of stem, sword shaped, 10-60cm in length and 1-2cm in width. Milky white, yellow white and red stripes on green leaves. Racemes on top of stems or branches. Berry, seeds 1-3.

## ECOLOGICAL DISTRIBUTION

Native to Madagascar. Distributed in China, Pakistan, etc.

## GROWTH HABIT

It likes high temperature, humid and sunny environment. It is drought resistant but not cold resistant. Suitable for growth is 20-30°C temperature .

旱，不耐寒；适生温度20~30℃。

## 用途、价值

多应用于园林绿化中，可以吸收空气中对人体有害的物质。花朵、叶片、根系均可药用，具有散热、止血等功效。

APPLICATION,VALUE

Mostly used in landscaping, it can absorb harmful substances in the air. Flowers, leaves and roots can be used medicinally, and have the functions of dissipating heat and stopping bleeding.

THE SAFETY OF OUR TEAM MEMBERS SHOULD ALWAYS COME BEFORE THE PRODUCTION OF OUR PARTS.

# 16 / 椰子

## 植物形态

常绿乔木。树高10~30m，茎粗壮，有环状叶痕，常有簇生小根。叶革质；线状披针形，长1~4m；裂片多数，线状披针形，外向折叠，长0.5~1m；叶柄粗壮，长1m以上。肉穗花序腋生，厚木质，老时脱落；花瓣3，卵状长圆形，长1~2cm。果卵球形或近球形，顶端具3棱，外果皮薄，中果皮厚纤维质，内果皮木质，基部有3孔，果腔含有胚乳；种子萌发时由孔穿出，种子1枚。花果期8~11月。

## 生态分布

原产于亚洲东南部、印度尼西亚至太平洋群岛。主要分布于中国、巴基斯坦、印度等地。

## 生长习性

喜光植物。在高温多雨、阳光充足、温差小且全年无霜的条件下生长发育良好；水分条件适宜为年降水量1200~2500mm，适生于海洋冲积土和河岸冲积土中。

## 用途、价值

椰肉可食用，椰纤维可制毛刷、地毯、缆绳等，椰壳可制成椰丝、椰砖作育苗基质、高级活性炭，树干可作建筑材料，根可入药，汁水可作饮料。还是

## PLANT MORPHOLOGY

Evergreen tree. Plant height 10-30m, with thick stems, annular leaf scars and often clustered small roots. Leaves leathery; linear lanceolate, 1-4m long; lobes numerous, linear lanceolate, outward folded, 0.5-1m long; petiole stout, more than 1m long. The fleshy inflorescence is axillary, thick woody, falling off when old; petals 3, ovate oblong, 1-2cm long. The fruit is ovoid or subglobose, with trigonous apex, thin epicarp, thick fibrous mesocarp, woody endocarp, with three pores at the base, and endosperm in the fruit cavity; the seed penetrates through the hole when it germinates. The flowering and fruiting period is from August to November.

## ECOLOGICAL DISTRIBUTION

Native to Southeast Asia, Indonesia to the Pacific Islands. Distributed in China, Pakistan, India, etc.

## GROWTH HABIT

Light-loving plants It grows well under the conditions of high temperature, rainy, sunny, small temperature difference and frost-free all year round; The water condition is suitable for the annual rainfall of 1200-2500mm, and it is suitable for the alluvial soil and the alluvial soil on the banks of the river.

## APPLICATION,VALUE

Coconut meat is edible, coconut fiber can be made into brushes, carpets, cables, etc, coconut shell can be made into shredded coconut and coconut bricks as nursery substrate and advanced activated carbon, trunk can be used as building materials, roots can be used as medicine, and juice can be

组织培养的良好促进剂。树形高大优美，是沙滩地区常见树种，具有较高的观赏价值，园林中应用也较广泛。是热带地区独特的可再生、绿色、环保型资源。

used as beverage. It is also a good promoter for tissue culture. The tree is tall and beautiful, and it is a common tree species in the beach area. It has high ornamental value and is widely used in gardens. It is a unique renewable, green and environment-friendly resource in tropical areas.

# 17 / 王棕

*Roystonea regia*
棕榈科 Arecaceae　大王椰属 *Roystonea*
别名：大王椰子、王椰、大王椰

## 植物形态

常绿乔木。单干，树高10~20m。茎基部膨大，近中部不规则膨大，向上部渐窄。叶羽状全裂，弓形，尾部下垂，长4~5m，叶轴每侧羽片多达250片，羽片4列，线状披针形，长约1m，宽3~5cm。花序长达1.5m，多分枝；佛焰苞2，开花前棒状，花后鞘状，花冠壶状，3裂至中部；花期3~4月。果近球形或倒卵形，长约2cm，直径约1cm，暗红或淡紫色，果期10月。种子歪卵形，1枚。

## PLANT MORPHOLOGY

Evergreen tree. Single trunk, plant height 10-20m. The base of the stem expanded irregularly near the middle and narrowed to the upper part. Leaves pinnatifid, arched, drooping tail, 4-5m in length, with 250 pinnae on each side of the leaf axis, 4 rows of pinnae, linear lanceolate, about 1m in length and 3-5cm in width. Inflorescence 1.5m in length, multi branched; spathe 2, club shaped before flowering, sheath shaped after anthesis, the corolla pot shaped, 3-lobed to the middle; flowering from March to April. Fruit nearly spherical or obovate, about 2cm in length and 1cm in diameter, dark red or light purple, and fruit period October. Seed 1, crooked ovate.

## 生态分布

原产古巴。分布于中国、巴基斯坦等地。

## 生长习性

喜高温高湿、阳光充足环境，不耐寒。抗风力强，较耐干旱，根系粗大发达。喜土层深厚肥沃的酸性土，不耐贫瘠。

## 用途、价值

可作为行道树，或植于高层建筑旁、水滨、草坪等处。种子可作饲料，其茎和叶可作建造材料。

## ECOLOGICAL DISTRIBUTION

Native to Cuba. Distributed in China, Pakistan, etc.

## GROWTH HABIT

Favoring high temperature and humidity, sunny environment, not cold-resistant. Strong wind resistance, drought resistance, thick and developed root system. Favoring deep and fertile acid soil, but not tolerate barren soil.

## APPLICATION,VALUE

Can be used as street trees, or planted beside high-rise buildings, waterfront, lawns and other places. Seeds can be used as feed and their stems and leaves can be used as building materials.

# 18 / 贝叶棕

*Corypha umbraculifera*
棕榈科 Arecaceae　贝叶棕属 *Corypha*
别名：长叶棕

## 植物形态

常绿乔木。树高18~25m，径可达90cm。茎密被环状叶痕，四回分枝；分枝花序从花苞的缝隙中抽出，分枝螺旋状着生小花枝。叶扇状深裂至中部，剑形，长0.6~1.0m；叶柄长2.5~3.0m，边缘具短齿，顶端延伸成下弯中肋状叶轴，长70~90cm。花乳白色，圆锥花序顶生、直立，长4~5m或更长。果球形，直径3~4cm，种子近球形或卵球形，直径1~2cm。

## 生态分布

原产印度、斯里兰卡等亚洲热带地区。分布于中国、巴基斯坦、缅甸、印度、斯里兰卡等地。

## 生长习性

喜高温高湿的生长环境，不耐寒，对土壤要求不严，喜光，较抗风。

## 用途、价值

树形美观，是很好的绿化观赏植物。叶片可作书写材料，汁液可制棕榈酒或醋或熬制成糖；幼嫩种仁可用糖浆煮成甜食；树干髓心可制淀粉供食用；幼株及根的汁液可治腹泻、热感冒。

## PLANT MORPHOLOGY

Evergreen tree. Plant height 18-25m, 90cm in diameter. Stem densely covered with annular leaf scars, four branches; branch inflorescence extracted from the gap of flower bud, and branch spiral with small flower branches. Leaves fan-shaped, deeply lobed to the middle, sword shaped, 0.6-1.0m in length; petioles 2.5-3.0m in length, with short teeth on the edge, and the top extending into a downward curved middle rib leaf axis, 70-90cm in length. Flowers milky white, panicles terminal, erect, 4-5m long or longer. Fruit globose, 3-4cm in diameter, seeds subglobose or ovoid, 1-2cm in diameter.

## ECOLOGICAL DISTRIBUTION

Native to tropical Asian countries such as India and Sri Lanka. Distributed in China, Pakistan, Myanmar, India, Sri Lanka, etc.

## GROWTH HABIT

It likes the growth environment of high temperature and humidity, is not cold-tolerant, does not have strict requirements on the soil, likes sunshine, and is more wind-resistant.

## APPLICATION,VALUE

The tree is beautiful, and it is a good greening ornamental plant. The leaves can be used as writing materials, and the juice can be made into palm wine or vinegar or boiled into sugar; Young kernels can be cooked into sweets with syrup (mature kernels are poisonous and inedible); The pith of trunk can be made into starch for eating; The sap of young plants and roots can cure diarrhea and cold.

# 19 / 桑

*Morus alba*

桑科 Moraceae　桑属 *Morus*

别名：家桑、蚕桑

## 植物形态

落叶乔木。树高1~15m。树皮灰色、浅纵裂。叶卵形或宽卵形，深绿色，基部圆或微心形，先端短尖，叶缘锯齿粗钝，有时缺裂，叶下面脉腋具簇生毛；长5~15cm，宽5~12cm。柔荑花序，花淡黄色，腋生或生于芽鳞腋内，与叶同出；总花梗长5~10mm，被柔毛，花期4~5月。聚花果卵状椭圆形，长1~3cm，红色至暗紫色；果期5~7月。

## 生态分布

原产中国。分布于朝鲜、日本、蒙古、巴基斯坦、俄罗斯、印度、越南等地。

## 生长习性

喜温暖湿润气候，稍耐阴。适生温度25~30℃。耐旱，耐瘠薄，不耐涝。对土壤的适应性强。

## 用途、价值

树皮可作纺织原料、造纸原料；叶为养蚕的主要饲料，桑叶可药用，可清肺明目，治风热感冒等。木材可制家具、乐器、雕刻等；桑椹可以酿酒。桑树树冠美观且能抗烟尘及有毒气体，适于城市、工矿区及农村"四旁"绿化；为良好的绿化及经济树种。

## PLANT MORPHOLOGY

Deciduous tree. Plant height 1-15m. Bark gray and slightly longitudinal. Leaves ovate or broadly ovate, dark green, base rounded or slightly cordate, apex short pointed, leaf margin serrate coarsely obtuse, sometimes dehiscent, vein axillary under leaf with tufted hairs 5-15cm in length, 5-12cm in width. Catkin, flowers pale yellow, axillary or born in axillary bud scales, with leaves out. Total pedicel 5-10mm in lengthand pilose. Flowering period from April to May. Fruit ovoid elliptic, 1-3cm in length, red to dark purple; fruiting period from May to July.

## ECOLOGICAL DISTRIBUTION

Native to China. Distributed in North Korea, Japan, Mongolia, Pakistan, Russia, India, Vietnam, etc.

## GROWTH HABIT

Favoring warm and humid climate, a little shade tolerant. The suitable growth temperature is 25-30℃. Tolerant to drought, barrenness, and intolerant to flooding. Strong adaptability to soil.

## APPLICATION, VALUE

Bark can be used as raw material for textile and paper making; leaves are the main feed for sericulture, and mulberry leaves can be used medicinally to clear the lungs and improve eyesight, and treat wind-heat and colds. Wood can be used to make furniture, musical instruments, carvings, etc.; mulberry can be used to make wine. The mulberry tree has a beautiful crown and is resistant to smoke and toxic gases. It is suitable for greening in cities, industrial and mining areas and rural areas; it is a good greening and economic tree species.

# 20 / 榄仁树

*Terminalia catappa*

使君子科 Combretaceae  榄仁树属 *Terminalia*

别名：山枇杷树、大叶榄仁

## 植物形态

常绿乔木。树高2~10m。树皮黑褐色，纵裂而剥落状；枝条近顶部密被棕黄色的茸毛，具密而明显的叶痕。叶互生，基部截形或窄心形，幼叶背面疏被软毛，边缘稀微波状；主脉粗壮，基部近叶柄处被茸毛。穗状花序，长10~20cm，花绿色或白色，无花瓣；萼筒杯状；花盘被白色粗毛；花期3~6月。果椭圆形，稍扁，具2纵棱，棱上具翅状窄边，成熟时青黑色，果皮木质，果期7~9月。种子矩圆形，1枚。

## PLANT MORPHOLOGY

Evergreen tree. Plant height 2-10m. Bark brown and black, longitudinally split and exfoliated; branches densely covered with brown hairs near the top, with dense and obvious leaf scars. Leaves alternate, base truncated or narrowly heart-shaped. underside of the young leaves sparsely soft and edges sparsely microwave-like; main veins strong, and base villi near the petiole. Spikes, 10-20cm in length, green or white flowers, no petals; calyx tube cup-shaped; disks covered with white shag; flowering period from March to June. fruit oval, slightly flattened, with 2 longitudinal edges, with wing-like narrow edges on the edges, blue-black when mature, with woody skin, fruiting period from July to September. Seed oblong, 1 piece.

## 生态分布

原产于马达加斯加、印度东部和安达曼群岛及马来半岛。分布于中国、巴基斯坦、马来西亚、越南等地。

## 生长习性

喜光，有较强的抗干旱和抗逆能力。适生于贫瘠的土壤，能够适应恶劣生境条件。

## 用途、价值

可净化空气、美化环境；是一种优良绿化树种。防风、防污染、净化水体，叶片能够在一定程度上降低水体pH值。木材可做家具、工艺品；是良好薪炭材；种仁可食用，根和未成熟的果壳可提取单宁；叶可作染料和养蚕饲料；树皮、叶子具有治疗支气管炎的药用价值。

## ECOLOGICAL DISTRIBUTION

Native to Madagascar, eastern India and the Andaman Islands and the Malay Peninsula. Distributed in China, Pakistan, Malaysia, Vietnam, etc.

## GROWTH HABIT

Favoring sunshine and has strong drought resistance and stress resistance. It is suitable for poor soil and can adapt to harsh habitat conditions.

## APPLICATION,VALUE

It can purify the air and beautify the environment; it is an excellent green tree species. Windproof, anti-pollution, and water purification, the blade can reduce the pH of the water to a certain extent. Wood can be used for furniture and handicrafts; it is a good firewood; the kernels are edible, the roots and immature husks can extract tannins; the leaves can be used as dyes and feed for sericulture; the bark and leaves have medicinal purposes such as treating bronchitis value.

# 21 / 丝葵

*Washingtonia filifera*

棕榈科 Arecaceae  丝葵属 *Washingtonia*

别名：老人葵、华盛顿棕榈树

## 植物形态

常绿乔木。树高10~20m。树干灰色。基部通常不膨大，向上为圆柱状，顶端稍细，可见明显纵裂和不太明显的环状叶痕。叶大型，直径达1.8m；叶柄约与叶片等长，老树叶柄下半部边缘具正三角形小刺。花序大型，弓状下垂；花萼管状钟形，裂片披针形渐尖，略具芒尖，花期7月。果实卵球形，直径约0.5cm，顶端具刚毛状的长约0.5cm的宿存花柱。种子卵形，长约7mm，宽5mm。

## PLANT MORPHOLOGY

Evergreen tree. Plant height 10-20m. trunk gray. base usually not expanded, and cylindrical upward, tip slightly thin, with obvious longitudinal crack and less obvious annular leaf trace. leaves large, up to 1.8m in diameter; petioles about the same length as the leaves, and lower part of petioles of old trees with regular triangular spines. Inflorescence large, arcuate, pendulous; calyx tubular, campanulate, lobes lanceolate, acuminate, slightly awned, flowering period in July. fruit ovoid, about 0.5cm in diameter, and top with a setiform persistent style about 0.5cm in length; seed ovate, about 7mm in length and 5mm in width.

## ECOLOGICAL DISTRIBUTION

Native to the United States. Introduced and

## 生态分布

原产美国。巴基斯坦、中国等地有引种栽培。

## 生长习性

喜温暖、湿润、向阳的环境，较耐寒，耐瘠薄。不宜在高温、高湿处栽培。对土壤要求不严。

## 用途、价值

宜栽植于庭园观赏，也可作行道树，是亚热带干旱地区的优良园林树种。叶片作工艺品，果实和顶芽可供食用，叶柄纤维可作牙签。

cultivated in Pakistan and China, etc.

### GROWTH HABIT

Favoring warm, humid, sunny environment, more resistant to cold, barren. It is not suitable for cultivation in high temperature and high humidity. The soil is not strict.

### APPLICATION,VALUE

It is suitable to be planted in gardens for viewing, and can also be used as a street tree. It is an excellent garden tree species in subtropical arid areas. The leaves are used as handicrafts, the fruits and top buds are edible, and the petiole fibers can be used as toothpicks.

## 22 / 长叶暗罗

*Polyalthia longifolia*
番荔枝科 Annonaceae　暗罗属 *Polyalthia*
别名：印度塔树、垂枝暗罗

### 植物形态

常绿乔木。树高6~18m，树皮纵向剥落；树体尖塔状，侧枝下垂。单叶互生，叶长披针形，嫩叶铜褐色，后渐变为深绿色，边缘波浪状，长12~30cm，下垂。聚伞花序簇生枝顶或枝条上部，花具长柄；花期5~6月。核果椭圆形，成熟时红黑色，聚生具柄，果期9~10月。种子有纵向凹槽。

### 生态分布

原产于印度、孟加拉国、东南亚等热带地区。分布于巴基斯坦、中国、斯里兰卡等地。

### 生长习性

喜高温，不耐霜冻、低温；喜光照充足而炎热的气候，不耐阴，适生温度22~32℃。喜排水良好的砂质土壤，有一定的抗盐碱性，不耐涝。

### 用途、价值

树形独特，叶面油亮，四季青翠，适合庭院美化或作为行道树。树皮可作为地毯、麻绳、制作衣服的材料；木材常用于制作鼓体、铅笔和盒子；可作为生物燃料。

## PLANT MORPHOLOGY

Evergreen tree. Plant height 6-18m, bark peeled longitudinally; tree body steeple and lateral branches drooping. Single leaf alternate, long lanceolate, tender leaves copper brown, dark green, edge wavy, 12-30cm in length, drooping. Cymes clustered at the top of branches or upper branches, flowers with long stalks; flowering period from May to June. Drupe elliptic, red black at maturity, aggregated stipitate, fruiting period from September to October; seeds longitudinally grooved.

## ECOLOGICAL DISTRIBUTION

Native to tropical regions such as India, Bangladesh and Southeast Asia. Distributed in Pakistan, China, Sri Lanka, etc.

## GROWTH HABIT

Favoring high temperature, frost resistance and low temperature; it is fond of a hot climate with sufficient sunlight, and it is not tolerant to shade. The suitable temperature for growth is about 22-32°C. It likes sandy soil with good drainage, and has certain salt alkali resistance and waterlogging resistance.

## APPLICATION,VALUE

The tree shape is unique, the leaves are shiny, and the four seasons are green, suitable for garden beautification or as a street tree. Bark can be used as a material for carpets, hemp rope, and clothes; wood is often used to make drums, pencils and boxes; it can be used as biofuel.

# 23 / 苏铁

*Cycas revoluta*
苏铁科 Cycadaceae　苏铁属 *Cycas*
别名：避火蕉、凤尾松、凤尾蕉、铁树

## 植物形态

常绿乔木。树高2~8m，皮干有明显螺旋状排列的菱形叶柄残痕。叶革质，羽状叶从茎的顶部生出，羽状叶呈倒卵状披针形，呈"V"字形伸展；叶柄长10~20cm，具刺6~18对。雄球花圆柱形，长30~60cm，直径8~15cm，密被灰黄色茸毛，花期6~7月。种子倒卵圆形或卵圆形，稍扁；红褐色或橘红色，长2~4cm，直径1~3cm；密生灰黄色短茸毛，两侧有棱脊，顶端有尖头；种子10月成熟。

## PLANT MORPHOLOGY

Evergreen tree. Plant height about 2-8m, trunk with obvious spiral rhombic petiole traces. leaves leathery, pinnate leaves growing from the top of the stem, pinnate leaves obovate lanceolate, extending in V shape; petiole 10-20cm in length, with 6-18 pairs of spines; male cone cylindrical, 30-60cm in length, 8-15cm in diameter, densely covered with grayish yellow villus, and its flowering period from June to July. seeds obovate or ovoid, slightly flat; reddish brown or orange red, 2-4cm in length and 1-3cm in diameter; densely grayish yellow short villi with ridges on both sides and sharp tips on the top; seeds mature in October.

## ECOLOGICAL DISTRIBUTION

Distributed in China, Pakistan, Japan,

## 生态分布

分布于中国、巴基斯坦、日本、菲律宾、印度尼西亚等地。

## 生长习性

喜光，喜暖热湿润的环境，不耐寒，耐干旱。喜肥沃和微酸性的土壤。

## 用途、价值

外形美观，常见观赏树种。茎内含淀粉，可供食用，叶含有丰富的氨基酸和糖，为人体提供充足的营养；苏铁叶、花、种子和根都可以用作药物，可治疗高血压等症，苏铁种子和茎顶部树心有毒。

Philippines, Indonesia, etc.

### GROWTH HABIT

Favoring sunshine, like warm and humid environment, not cold and drought resistant. Like fertile and slightly acidic soil.

### APPLICATION,VALUE

Beautiful appearance, common ornamental tree species. The stem contains starch, which is edible. The leaves are rich in amino acids and sugars, which provide sufficient nutrition for the human body. Cycad leaves, flowers, seeds and roots can all be used as medicines to treat hypertension and other diseases. Seeds and the heart of the stem tops are poisonous.

# 24 / 柑橘

*Citrus reticulata*
芸香科 Rutaceae　柑橘属 *Citrus*
别名：蜜橘、黄橘、红橘

## 植物形态

常绿小乔木。枝扩展或略下垂，具少量枝刺。单生复叶，披针形、椭圆形或阔卵形，顶端常有凹口，叶缘上半段通常有钝或圆裂齿，少全缘。花黄白色或带淡红色；单生或2~3朵簇生；扁圆形至近圆球形；花期4~5月。果皮薄而光滑，或厚而粗糙；橘络呈网状，通常柔嫩，中心柱大而常空；汁胞通常纺锤形，短而膨大；果肉酸或甜，果期10~12月。种子卵形。

## 生态分布

原产中国。分布于巴基斯坦、巴西、美国、伊朗、印度、意大利等地。

## 生长习性

适生温度12~37℃，适生于湿润的土壤环境，耐阴性较强，以pH值5~6为最适宜。

## 用途、价值

果实食用价值高，橘皮有促进胃肠道消化、消炎、抑菌等作用。花、叶、果皮都是提取香精的优质原料和高级饮料、点心的调味剂。果皮主要用于食品、制药，可用作制糖加工原料；果渣可提取色素，果胶可作饲料。也可作庭园观赏植物。

## PLANT MORPHOLOGY

Small evergreen tree. Branches extended or slightly pendulous, with a small number of spines. Leaves single compound, lanceolate, elliptic or broadly ovate, often with a notch at the tip, the upper part of the leaf margin usually has obtuse or round cleft teeth, less entire. Flowers yellow white or reddish; solitary or 2-3 clustered; oblate to sub globose; flowering period from April to May. Pericarp thin and smooth, or thick and rough; orange reticulate, usually soft, central column large and often empty; juice cells usually spindle shaped, short and expanded; flesh sour or sweet, and fruiting period from October to December. Seeds ovate.

## ECOLOGICAL DISTRIBUTION

Native to China. Distributed in Pakistan, Brazil, the United States, Iran, India, Italy, etc.

## GROWTH HABIT

Suitable for 12-37°C temperature, suitable for moist soil environment, strong adaptability to negative, most suitable for pH 5-6.

## APPLICATION,VALUE

Fruit edible, orange peel effective in promote gastrointestinal digestion, anti-inflammatory, antibacterial and other functions. Flowers, leaves and pericarp use as high quality raw materials, premium drinks and dim sum flavoring agents. Peel mainly used in food and pharmaceutical industry, and raw material for sugar processing; pigment extracted from fruit residue, and pectin used as feed and garden ornamental plants.

# 25 / 乌柏

*Triadica sebifera*

大戟科 Euphorbiaceae 乌柏属 *Triadica*
别名：腊子树、柏子树、木子树

## 植物形态

落叶乔木。树高3~15m，树皮有纵裂纹，枝条与叶片具乳状汁液体。叶互生，纸质，菱形，顶端渐尖，基部阔楔形或钝，全缘。花单性，总状花序，聚集成顶生；基部两侧各具一近肾形的腺体，每一苞片内具10~15朵花；花梗粗壮，花期4~8月。果梨状球形，成熟时黑色，直径1~2cm。种子扁球形，黑色，长约1cm，宽约6mm，外被白色、蜡质的假种皮，种子3枚。

## 生态分布

分布于中国、巴基斯坦、印度、美

## PLANT MORPHOLOGY

Deciduous tree. Plant height 3-15m, bark longitudinal cracks, branches and leaves have milky juice. Leaves alternate, papery, rhombic, apex acuminate, base broadly cuneate or obtuse, entire. Flowers unisexual, racemose, aggregated and terminal; each side of the base has a subrenal gland with 10-15 flowers in each bract; pedicel robust, flowering period from April to August. Fruit pear shaped, black at maturity, 1-2cm in diameter. Seeds oblate, black, about 1cm in length, 6mm in width, covered with white, waxy aril, and 3 seeds.

## ECOLOGICAL DISTRIBUTION

Distributed in China, Pakistan, India, the United States, etc.

国等地。

## 生长习性

喜光，适生温度15℃，年降水量在750mm以上地区。较耐干旱；对土壤适应性较强。抗风、抗毒气。

## 用途、价值

常用于园林绿化中。根皮、树皮、叶均可入药，用于杀虫等。种子可供制高级香皂、蜡烛、蜡纸等。种仁榨油可作涂料、油漆、油墨等。木材可作家具和雕刻等。叶可作染料，假种皮可制肥皂等。

## GROWTH HABIT

Light-favored, suitable for 15°C temperature, annual more than 750mm rainfall. drought resistance and strong adaptability to soil. Anti wind, anti poison gas.

## APPLICATION,VALUE

Often use in landscaping. Root bark, bark, and leaves use as medicine to kill insects. Seeds use to make high-grade soap, candles, wax paper, etc. Kernel oil use as coating, paint, ink, etc. Wood use for furniture and carvings. Leaves use as dyes, and arils use to make soap.

## 26 / 榕树

## 植物形态

常绿乔木。树高10~25m。树皮深灰色，老树常有锈褐色气生根。叶薄革质，椭圆形，先端钝尖，基部楔形，全缘，托叶披针形，长约1cm。隐头花序，雄花、雌花同生于一榕果内，花间具少数刚毛；花期5~6月。果成对腋生或生于已落叶枝叶腋；扁球形，直径约1cm；成熟时黄色或微红色。

## 生态分布

分布于中国、斯里兰卡、印度、巴基斯坦、泰国、越南、马来西亚、菲律宾、日本、巴布亚新几内亚等地。

## PLANT MORPHOLOGY

Evergreen tree. Plant height 10-25m; bark dark gray, old trees often having rusty brown aerial roots. Leaves thin leathery, oval, blunt tip, cuneate base, entire, stipules lanceolate, about 1cm in length. Inflorescence hypanthodium, male and female flowers born in a banyan fruit, with a few bristles between flowers; flowering period from May to June. Fruit axillary in pairs or born in the axils of deciduous branches; oblate, about 1cm in diameter; yellow or reddish when mature.

## ECOLOGICAL DISTRIBUTION

Distributed in China, Sri Lanka, India, Pakistan, Thailand, Vietnam, Malaysia, Philippines, Japan, Papua New Guinea, etc.

## 生长习性

喜温暖湿润气候，喜阳光充足的生长环境，忌暴晒。不耐寒，不耐旱，较耐水湿，适应性强，对土壤要求不严，喜疏松肥沃的酸性土。

## 用途、价值

在园林应用中，常用作行道树，美化庭园，露地栽培，可制作成盆景，亦可作为孤植树观赏。树皮纤维可制渔网和人造棉；气生根、树皮和叶芽可入药，有止咳、清热利湿、止泻等功效。

## GROWTH HABIT

Favoring warm and humid climate, light-favored, avoid exposure. Unhardy, not drought resistance, water and humidity resistance, strong adaptability, not strict to the soil quality, favoring loose and fertile acid soil.

## APPLICATION, VALUE

In garden applications, often used as street trees, beautify gardens, cultivate in open fields, making into bonsai, or as a solitary tree for viewing. Bark fiber used to make fishing nets and artificial cotton; aerial roots, bark and leaf buds applicable in medicine, effective in relieving cough, clearing heat and removing dampness, and relieving diarrhea.

# 27 / 高山榕

*Ficus altissima*
桑科 Moraceae　榕属 *Ficus*
别名：大青树、大叶榕

## 植物形态

　　高大乔木。树高25~30m。叶厚革质，广卵状椭圆形，长10~20cm，宽5~10cm，叶面光滑，叶柄粗壮，长1~4cm；托叶厚革质，外面被灰色绢丝状毛。隐头花序，雌雄同序，花期3~4月。果成对腋生，椭圆状卵圆形，直径1~3cm，红色或带黄色，顶部脐状，基生苞片短宽，脱落后环状；果期5~7月。

## 生态分布

　　原产中国。分布于尼泊尔、印度、不丹、巴基斯坦、缅甸、越南、泰国、

## PLANT MORPHOLOGY

　　Evergreen tree. Plant height 25-30m, leaves thick leathery, broad oval-shaped, 10-20cm in length, 5-10cm in width, smooth leaf surface, stout petiole, 1-4cm in length; stipules thick leathery and covered with gray silk silk hairy. Hypanthodium, androgynous, flower period from March to April. Fruit axillary in pairs, elliptic ovoid, 1-3cm in diameter, red or yellowish, umbilicus at the top, basal bracts short and wide, ring-shaped after shedding; fruiting period from May to July.

## ECOLOGICAL DISTRIBUTION

　　Native to China. Distributed in Nepal, India, Bhutan, Pakistan, Myanmar, Vietnam, Thailand, Malaysia, Indonesia, etc.

马来西亚、印度尼西亚等地。

## 生长习性

喜阳光充足、高温的生长环境。耐干旱。对土壤要求不严，在肥沃、湿润、排水良好的土壤中生长良好。

## 用途、价值

常作行道树和孤植树；抗风、抗大气污染，是极好的城市绿化树种，适合用作园景树和遮阴树，适合在室内长期陈设。所形成的独树成林的景观具有重要的观赏价值。

## GROWTH HABIT

Light-favored, high-temperature growth environment. drought Resistance. Not strict to the soil quality, growing well in fertile, moist, and well-drained soil.

## APPLICATION, VALUE

Often used as street trees and solitary trees; resistant to wind and air pollution, excellent urban greening tree species. suitable for gardening and shade trees, and suitable for long-term indoor furnishings. The resulting landscape of single trees and forests with important ornamental value.

# 28 / 刺黄果

*Carissa carandas*

夹竹桃科 Apocynaceae 假虎刺属 *Carissa*
别名：瓜子金、林那果

## 植物形态

小乔木。树高1~5m。刺长5cm；叶对生，革质，长椭圆形，暗绿色；叶近边缘网结。花白色或淡玫瑰色，聚伞花序顶生，常具3花，花序梗长1~3cm；裂片披针形，被微柔毛，花冠筒长2cm，内面被微柔毛；花萼基部内面具多数腺体；花期3~6月。浆果卵球形，紫红色，直径1~2cm。种子长圆形，呈压扁状；果期7~12月。

## 生态分布

分布于中国、巴基斯坦、印度、斯里兰卡、缅甸和印度尼西亚等地。

## PLANT MORPHOLOGY

Small tree. Plant height 1-5m. Spines 5cm in length; Leaves opposite, leathery, oblong, dark green; leaves knotted near the edge. Flowers white or pale rosy. Inflorescence cymes terminal, often with 3 flowers, peduncle 1-3cm in length; lobes lanceolate, puberulent, corolla tube 2cm in length, inner surface puberulent; calyx base Most glands in the inner mask; flowering period from March to June. Berry ovoid, purple-red, 1-2cm in diameter; seeds oblong, squashed; fruiting period from July to December.

## ECOLOGICAL DISTRIBUTION

Distributed in China, Pakistan, India, Sri Lanka, Myanmar and Indonesia, etc.

## 生长习性

喜高温，日照需充足，耐干旱，越冬温度5℃以上；栽培处排水需良好，要求疏松肥沃土壤。

## 用途、价值

果实可生食，也可作糕饼馅及果酱等。刺长而锐利，可作围篱；叶色浓绿而光亮，常作花篱，盆栽可装饰庭院、客厅、厅堂。

## GROWTH HABIT

Favoring high temperature, sufficient sunshine, drought resistance and overwintering temperature above 5℃; good the drainage of cultivation site and loose, requiring fertile soil.

## APPLICATION,VALUE

Fruit edible, used for fillings and jams. thorns long and sharp, used as fences; leaves dark green and bright, often used as flower fences. Potted plants applicable in decorate courtyards, guest rooms and halls.

# 29 / 珍珠金合欢

*Acacia podalyriifolia*
豆科 Leguminosae　相思树属 *Acacia*
别名：银叶金合欢

## 植物形态

常绿小乔木。树高2~5m。树干分枝低，主干不明显，树皮灰绿色，具灰色条纹。叶状柄宽卵形或椭圆形，被白毛与白粉，呈灰绿至银白色，基部圆形。头状花序，多个排列成总状花序，花黄色；花期3~6月。荚果扁平，褐色，果期7~11月。

## 生态分布

原产热带美洲。分布于中国、巴基斯坦等地。

## PLANT MORPHOLOGY

Small evergreen tree. Plant height 2-5m. Trunk low branches, main trunk not obvious, bark gray-green with gray stripes. Petiole broadly oval or elliptical, covered with white hairs and white powder, grayish green to silvery white, base round. Inflorescence heads, multiple arranged into racemes, with yellow flowers; flowering period from March to June. pods flat, brown, and fruiting period from July to November.

## ECOLOGICAL DISTRIBUTION

Native to tropical America. Distributed in China, Pakistan, etc.

## GROWTH HABIT

## 生长习性

喜光，喜温，耐旱，适生于排水良好的土壤。

## 用途、价值

常用于园林绿化。木材坚硬，可为贵重器材；根及荚果含单宁，可作染料，入药能收敛、清热；花可提取香精；茎流出的树脂可供美工用及药用。

Thermophilic, light-favored, drought resistance, suitable for soil with good drainage.

### APPLICATION,VALUE

Often used for landscaping. Wood hard, valuable equipment; root and pod contain tannin, dyestuff, medicine astringent, heat clearing; flower extractable flavors; the resin out of stems applicable in beauty and medicine.

# 30 / 番木瓜

*Carica papaya*
番木瓜科 Caricaceae　番木瓜属 *Carica*
别名：树冬瓜、番瓜、万寿果、木瓜

## 植物形态

常绿小乔木。树高2~10m。具乳汁。叶大而聚生茎顶，近盾形，直径达60cm；每裂片羽状分裂，5~9深裂；叶柄中空，长0.5~1.0m。花冠乳黄色，披针形，下垂，无梗；花单生或由数朵排列成伞房花序，乳黄色或黄白色。浆果肉质，倒卵状、长球形、梨形或近球形，长10~30cm或更长；成熟时橙黄或黄色；果肉柔软多汁，味香甜。种子多数，卵球形，成熟时黑色，花果期全年。

## PLANT MORPHOLOGY

Small evergreen tree. Plant height 2-10m. Leaves large and congregate on the top of the stem, nearly shield-shaped, 60cm in diameter; each lobe pinnately divided and 5-9 deeply divided; petiole hollow, 0.5-1.0m in length. Corolla creamy yellow, lanceolate, drooping, sessile; flowers solitary or arranged into corymbs, creamy yellow or yellowish white. Berry fleshy, obovate, oblong, pear-shaped or sub-globose, 10-30cm or longer in length; orange or yellow when mature; soft and juicy, sweet, with many seeds, ovoid, black when mature, flowering and fruiting period whole year.

## ECOLOGICAL DISTRIBUTION

Native to tropical America. Distributed in China, Pakistan, Malaysia, Philippines, Thailand,

## 生态分布

原产于热带美洲。主要分布于中国、巴基斯坦、马来西亚、菲律宾、泰国、越南、缅甸、印度尼西亚、印度、斯里兰卡、美国、古巴以及澳大利亚。

# 生长习性

喜阳光充足、高温多湿的生长环境，不耐寒，适生年均气温22~25℃，耐湿，抗旱。对土壤适应性较强。

## 用途、价值

药食同源。果实鲜美兼具食疗作用，含有胡萝卜素和丰富的维生素C，有很强的抗氧化能力，可增强人体免疫力，帮助机体抵抗病毒。果实加工产品可作化工、化妆品原料、饲料或添加剂，也可加工成蜜饯、果酱、罐头等。种子可榨油。具有较高的园林绿化观赏价值。

Vietnam, Myanmar, Indonesia, India and Sri Lanka, the United States, Cuba and Australia.

**GROWTH HABIT**

Favoring sunny, heat and humid environment. Not cold-resistant. suitable for temperature is 22-25℃. Moisture and drought resistance. Strong adaptability to soil.

**APPLICATION,VALUE**

Medicinal and edible. The fruit is delicious and has therapeutic effect. It contains carotene and rich vitamin C. It has strong antioxidant capacity, enhances human immunity and helps the body resist virus. Fruit processing products can be used as chemical, cosmetic raw materials, feed or additives. It can also be processed into preserves, jam, cans, etc. The seed can squeeze oil. It has greening and ornamental value.

# 31 / 腊肠树

*Cassia fistula*

豆科 Leguminosae　腊肠树属 *Cassia*

别名：阿勃勒、牛角树、波斯皂荚

## 植物形态

落叶乔木。树高2~15m；树皮灰色至暗褐色。羽状复叶，长30~40cm；小叶3~4对，长5~15cm，全缘，幼时被微柔毛；叶轴与叶柄无翅。总状花序，长达30cm，下垂；直径3~4cm，花黄色；花梗长3~5cm，萼片开花时反折。荚果圆柱形，长30~70cm，宽2~4cm，幼时绿色，成熟时黑褐色，不开裂，有3条槽纹。种子30~100枚。

## 生态分布

原产印度、缅甸和斯里兰卡。分布于中国、巴基斯坦、印度、泰国等地。

## PLANT MORPHOLOGY

Deciduous tree. Plant height 2-15m; bark gray to dark brown. Pinnate compound leaves, 30-40cm in length; 3-4 pairs of small leaves, 5-15cm in length, entire, puberulent when young; leaf axis and petioles wingless. Racemes, up to 300mm in length, drooping; about 30-40mm in diameter, yellow flowers; pedicels 30-50mm in length, and sepals reflexed when they bloom. Pod cylindrical, 30-70cm in length, 2-4cm in width, green when young, dark brown when mature, without cracking, with 3 flutes; 30-100 seeds.

## ECOLOGICAL DISTRIBUTION

Native to India, Myanmar and Sri Lanka. Distributed in China, Pakistan, India, Thailand, etc.

## 生长习性

喜光、也耐阴，耐湿，忌积水；抗风性强。适生温度20~32℃，能耐最低温度为-3℃。对土壤适应性较强。

## 用途、价值

是常见的园林绿化观赏树木。树皮可作染料；根、树皮、种子均可入药；木材可作支柱、桥梁、农具等用材。

## GROWTH HABIT

Light-favored, shade and humidity resistance, water repellency, strong wind resistance. Suitable for temperature is 20-32℃, and the lowest temperature for tolerance is -3℃. Strong adaptability to soil.

## APPLICATION,VALUE

It is a common landscaping ornamental tree. Bark can be used as dye; roots, bark, seeds can be used as medicine; wood can be used as pillars, bridges, agricultural tools and other materials.

# 32 / 菩提树

*Ficus religiosa*
桑科 Moraceae 榕属 *Ficus*
别名：思维树、菩提榕、觉树、娑罗双树

## 植物形态

常绿乔木。树高2~25m；小枝灰褐色，幼时被微柔毛。叶革质，三角状卵形，深绿色，先端尾尖长2~5cm，全缘或波状；基生三出脉，侧脉5~7对；叶柄与叶片等长或长于叶片；托叶卵形。隐头花序，雄花、瘿花和雌花生于同一榕果内壁；花期3~4月。果球形或扁球形，直径1~2cm，熟时红色；果期5~6月。

## 生态分布

分布于中国、巴基斯坦、日本、马来西亚、泰国、越南、不丹、印度、尼泊尔等地。

## 生长习性

喜光，喜高温高湿，不耐寒。对土壤要求不严，抗污染能力强。

## 用途、价值

用于庭院街道、污染区的绿化树种。枝干上流出的乳状液汁，可提取硬性橡胶；枝叶可作饲料；木材适宜作砧板、包装箱板和纤维板原料。具有药用价值，是治疗哮喘、糖尿病、癫痫等疾病的传统中医药。

## PLANT MORPHOLOGY

Evergreen tree. Plant height 2-25m; branchlets grayish brown, puberulent when young. Leaves leathery, triangular ovate, dark green, apex 2-5cm in length, entire or undulate; basal three veins, lateral veins 5-7 pairs; petiole and leaf blade equal length or longer than leaves; stipules ovate. Hypanthodium, male flowers, gall flowers and female flowers are borne on the inner wall of the same fig. Flowering period from March to April. Fruit spherical or oblate, 1-2cm in diameter, red at maturity. Fruiting period from May to June.

## ECOLOGICAL DISTRIBUTION

Distributed in China, Pakistan, Japan, Malaysia, Thailand, Vietnam, Bhutan, Sikkim, Nepal, etc.

## GROWTH HABIT

Favoring light, heat and humidity, not cold-resistant. Not strict to the soil quality, and strong anti-pollution ability.

## APPLICATION,VALUE

Used for greening trees in courtyard and polluted area. The sap flowing out of the branches can be made of hard rubber; the branches and leaves can be used as feed; wood is suitable for cutting boards, packaging box boards and fiberboard materials. It has medicinal value and is a traditional Chinese medicine for treating asthma, diabetes, epilepsy and other diseases.

# 33 / 灰莉

*Fagraea ceilanica*
龙胆科 Gentianaceae　灰莉属 *Fagraea*
别名：华灰莉、非洲茉莉、华灰莉木

## 植物形态

常绿乔木。树高2~10m。树皮灰色，老枝上有凸起的叶痕。叶片稍肉质，椭圆形、卵形或长圆形，长5~25cm，宽2~10cm，深绿色；托叶呈鳞片状。聚伞花序；花萼肉质；花冠漏斗状，白色，上部内侧有突起的花纹；花期4~8月。浆果，卵状或近圆球状，长3~5cm，直径2~4cm，淡绿色；果期7月至翌年3月。种子椭圆状肾形，长3~4mm。

## 生态分布

分布于中国、巴基斯坦、印度、泰

## PLANT MORPHOLOGY

Evergreen tree. Plant height 2-10m. Bark gray, with protruding leaf marks on the old branches. Leaves slightly fleshy, elliptic, ovate or oblong, 5-25cm in length and 2-10cm in width, dark green; stipules scaly. Cyme; calyx fleshy; corolla funnel-shaped, white, with protuberant pattern on the upper inner side; flowering period from April to August. Berry, ovoid or subglobose, 3-5cm in length, 2-4cm in diameter, light green; fruiting period from July to March of the next year. Seeds elliptic reniform, 3-4mm in length.

## ECOLOGICAL DISTRIBUTION

Distributed in China, Pakistan, India, Thailand, Vietnam, Philippines, Malaysia, etc.

国、越南、菲律宾、马来西亚等地。

## 生长习性

喜阳光，耐旱，耐阴，耐寒力强。对土壤要求不严，适应性强。

## 用途、价值

是庭园观赏植物；防风沙，抗污染能力强，适合于道路隔离带、交通主干道路、林带等地的绿化。枝、叶、花能产生挥发性油类，具有杀菌作用。

## GROWTH HABIT

Light-favored, drought and shade resistance, strong cold resistance. Not strict to the soil quality and with strong adaptability.

## APPLICATION,VALUE

Garden ornamental plant, wind and sand proof, strong anti pollution ability, suitable for the greening of road separation belt, traffic trunk road, forest belt and other places. Branches, leaves and flowers can produce volatile oil, which has bactericidal effect.

# 34 / 木麻黄

*Casuarina equisetifolia*
木麻黄科 Casuarinaceae　木麻黄属 *Casuarina*
别名：短枝木麻黄、马尾树、驳骨树

## 植物形态

常绿乔木。树高1.5~40m。树皮暗褐色，纤维质，呈窄长条片剥落，内皮深红色；枝红褐色，节密集；小枝灰绿色，纤细，柔软下垂，节间短，节易折断。鳞片状叶每轮常具7枚，披针形或三角形，长1~3mm，紧贴小枝，鳞片淡绿色，近透明。雄花序棒状圆柱形，长1~4cm，雌花序顶生于侧生短枝；花期4~5月。球果状果序椭圆形，长1.5~2.5cm，幼时具灰绿色或黄褐色茸毛。小坚果连翅长0.5~1.0cm，宽0.2~0.3cm；果期7~10月。

## PLANT MORPHOLOGY

Evergreen tree. Plant height 1.5-40m. Bark dark brown, fibrous, peeling off into narrow strips, and the endothelium dark red; branches reddish brown with dense nodes; branchlets gray-green, slender, soft and drooping, with short internodes, and nodes easily broken. Scaly leaves often having 7 per wheel, lanceolate or triangular, 1-3mm in length, close to the branchlets, and the scales pale green and nearly transparent. Male inflorescence rod-shaped and cylindrical, 10-40mm in length, and female inflorescence terminally born on lateral short branches, flowering period from April to May. Cone-shaped fruit sequence oval, 15-25mm in length, with gray-green or yellow-brown hairs when young. Small nut with wings 5-10mm in length and 2-3mm in width; fruiting period from July to October.

## 生态分布

原产澳大利亚。分布于中国、美国、巴基斯坦等地。

## 生长习性

喜热喜光，喜高温多雨的环境。耐干旱、贫瘠，抗盐渍，耐潮湿，不耐寒；适生于pH6.0~8.0的滨海沙土，对立地条件要求不高。

## 用途、价值

生长迅速，萌芽力强，根系深广，具耐旱、抗风沙和耐盐碱的特性，是热带海岸防风固沙的优良先锋树种；在城市及郊区亦可作行道树、防护林或绿篱。木材经防腐防虫处理后，可作枕木、船底板及建筑用材；为栲胶原料和医药上收敛剂；枝叶药用；幼嫩枝叶可为牲畜饲料。

## ECOLOGICAL DISTRIBUTION

Native to Australia. Distributed in China, the United States, Pakistan, etc.

## GROWTH HABIT

Favoring heat, light and rainy environment. Drought, barren, saline-alkail and humidity resistance, but cold-resistant; Favored growing in coastal sandy soil with pH 6.0-8.0, and not strict to the site conditions.

## APPLICATION,VALUE

Rapid growth, strong budding power, deep and wide root system. Drought, wind-borne sand and saline-alkali resistance. It is an excellent pioneer tree species for wind and sand fixation on tropical coasts; it can also be used as street trees, shelter forests or hedges in cities and suburbs. After the wood is treated with antiseptic and pest control, it can be used as sleepers, ship bottom plates and construction materials; as raw materials for tannin extracts and medical astringents; branches and leaves are used for medicine; young branches and leaves can be used as livestock feed.

# 35 / 桉

*Eucalyptus robusta*
桃金娘科 Caricaceae　桉属 *Carica*
别名：桉树、大叶桉、大叶尤加利

## 植物形态

常绿乔木。树高2~20m。树皮深褐色，不剥落，有不规则斜裂沟，幼枝有棱。叶厚革质，卵形或卵状披针形，基部偏斜，两面均有腺点，侧脉多而明显，长8~20cm，宽3~7cm。伞形花序，花白色带黄色，花梗短粗而扁平；花期4~9月。蒴果卵状壶形，长1~2cm；果瓣3~4片，藏于萼筒内。

## 生态分布

原产澳大利亚。分布于中国、巴基斯坦、印度尼西亚等地。

## PLANT MORPHOLOGY

Evergreen tree. Plant height 2-20m. Bark dark brown, not flaking, with irregular oblique furrows, young branches with ribs. leaves thick and leathery, ovate or ovate lanceolate, with oblique base, glandular spots on both sides, and numerous and obvious lateral veins, 8-20cm in length and 3-7cm in width. Umbel shaped inflorescence, white and yellow flowers, pedicel short thick and flat, flowering from April to September. Capsule ovate, pot shaped, 1-2cm in length; fruit petals 3-4, hidden in the calyx tube.

## ECOLOGICAL DISTRIBUTION

Native to Australia. Distributed in China, Pakistan, Indonesia, etc.

## 生长习性

喜光，喜湿，耐旱，耐热，畏寒。适生于酸性的红壤、黄壤和土层深厚的冲积土。

## 用途、价值

主根深、抗风力强、萌芽更新能力强、抗旱能力强，宜作行道树、防风固沙林和园林绿化树种。可造牛皮纸和打印纸，木材抗腐能力强，可用作枕木、矿柱、电杆等，木材可作燃料，叶可作饲料；树根可食用，鲜叶可提油，亦可作药用原料。

## GROWTH HABIT

Light-favored, wet-loving, drought resistant, heat-resistant, and chilly. It is suitable for acid red soil, yellow soil and deep alluvial soil.

## APPLICATION,VALUE

With deep taproot, strong wind resistance, strong germination and regeneration ability and strong drought resistance, it is suitable for street trees, windbreak and sand fixation forests and landscaping trees. It can be used for kraft paper and printing paper. Wood has strong corrosion resistance and can be used as sleepers, pillars, poles, etc. Wood can be used as fuel and leaves can be used as feed. Roots are edible, fresh leaves can be used to extract oil, and can also be used as medicinal raw materials.

**36 / 刺桐**

## 植物形态

乔落叶乔木。树高2~20m。树皮灰褐色，分枝有圆锥形黑色皮刺。羽状复叶具3小叶；小叶宽卵形或菱状卵形，长15~20cm，先端渐尖而钝，基部宽楔形或平截；叶柄长10~15cm。总状花序顶生，长10~15cm，宽5~10cm；花红色，旗瓣椭圆形，龙骨瓣2片分离，与翼瓣近等长；花萼佛焰苞状，偏裂，分裂到基部。荚果圆柱形，黑色，长15~30cm，直径2~3cm。种子肾形，暗红色，1~8枚。

## 生态分布

原产印度至大洋洲海岸林中。分布于巴基斯坦、中国、马来西亚、印度尼西亚、柬埔寨、老挝、越南等地。

## 生长习性

喜温暖湿润、光照充足的环境，耐旱、耐湿、不耐寒，适生于肥沃而排水良好的砂壤土。

## 用途、价值

是优良的行道树，适用于园林绿化，适合单植于草地或建筑物旁。树皮或根皮入药，有祛风湿、舒筋通络、治疗跌打损伤等作用。

## PLANT MORPHOLOGY

Deciduous tree. Plant height 2-20m. Bark grayish brown, and branches having conical black thorns. Pinnate compound leaf 3 leaflets; Leaflets broadly ovate or rhomboid-ovate, 15-20cm in length, acuminate and obtuse at the apex, broadly wedge-shaped or truncated at the base; petiole 10-15cm in length. Raceme terminal, 10-15cm in length and 5-10cm in width; flower red, te flag petal elliptical, and keel petals 2 separated, nearly as long as the wing petals; e calyx bud-shaped, split, split to base. Pods, cylindrical, black, 15-30cm in length and 2-3cm in diameter. seeds kidney-shaped, dark red, 1-8 pieces.

## ECOLOGICAL DISTRIBUTION

Native to the coastal forests from India to Oceania. Distributed in Pakistan, China, Malaysia, Indonesia, Cambodia, Laos, Vietnam, etc.

## GROWTH HABIT

Favoring warm and humid environment with sufficient sunlight, drought-tolerant, moisture-tolerant, and cold-tolerant, suitable for fertile sandy loam with good drainage.

## APPLICATION,VALUE

It is an excellent street tree, suitable for landscaping, suitable for single planting in grassland or beside buildings. Bark or root bark used as medicine has the functions of expelling wind and dampness, relaxing muscles and tendons and dredging collaterals, and treating traumatic injuries.

# 37 / 蓝花楹

## 植物形态

　　落叶大乔木。树高5~20m。叶对生，二回羽状复叶，羽片通常在16对以上，小叶8~24对，椭圆状披针形，长6~12mm，宽2~7mm，顶端急尖，基部楔形。花蓝色，花冠筒细长，上部膨大，下部微弯，花萼筒状；花期5~6月。蒴果木质，扁卵圆形，淡褐色，长宽均约5cm。

## 生态分布

　　原产南美洲巴西、玻利维亚、阿根廷。分布于中国、巴基斯坦等地。

## 生长习性

　　喜温暖、湿润、阳光充足的环境，不耐霜雪；对土壤条件要求不严。

## 用途、价值

　　可作行道树、遮阴树和风景树。是一种重要的木本花卉。可用于制作家具，也是优良的造纸树种。

## PLANT MORPHOLOGY

Large deciduous tree. Plant height 5-20m. Leaves opposite, 2-pinnate compound leaves, pinnae usually more than 16 pairs, leaflets 8-24 pairs, elliptic lanceolate, 6-12mm in length, 2-7mm in width, apex acute, base cuneate. flower blue, corolla tube slender, upper part enlarged, lower part slightly curved, calyx tubular, and flowering period May-June. Capsule woody, flat ovoid, light brown, approximately 5cm in length and width.

## ECOLOGICAL DISTRIBUTION

Native to Brazil, Bolivia and Argentina in South America. Distributed in China, Pakistan, etc.

## GROWTH HABIT

Favoring warm and humid, sunny environment, frost and snow intolerance; not strict to the soil quality.

## APPLICATION,VALUE

Can be used as street trees, shade trees and landscape trees. It is an important woody flower. It can be used to make furniture and is also an excellent paper tree species.

# 38 / 杧果

*Mangifera indica*
漆树科 Anacardiaceae　杧果属 *Mangifera*
别名：檬果、芒果、蜜望子、望果

## 植物形态

常绿乔木。树高5~20m。树皮灰褐色，小枝褐色，无毛。叶长圆形或圆状披针形，长12~30cm，先端渐尖，侧脉20~25对。圆锥花序，花多且密，黄色或淡黄色；具总梗，被黄色微柔毛；萼片被微柔毛；花瓣具3~5突起脉纹。核果肾形，长5~15cm，宽3~5cm，成熟时黄色，中果皮肉质，肥厚，鲜黄色，味甜，果核坚硬。

## 生态分布

原产印度。分布于中国、巴基斯坦、孟加拉国、马来西亚等地。

## PLANT MORPHOLOGY

Evergreen tree. Plant height 5-20m. Bark gray brown, branchlets brown, glabrous. Leaves oblong or oblong lanceolate, 12-30cm in length, apex acuminate, lateral veins 20-25 pairs. Panicle, many and dense flowers, yellow or light yellow; peduncle, yellow puberulent; sepals puberulent; petals with 3-5 protuberant veins. The stone fruit is kidney shaped, 5-15cm in length and 3-5cm in width. Yellow when mature, fleshy in mesocarp, thick, bright yellow, sweet and hard in stone.

## ECOLOGICAL DISTRIBUTION

Native to India. Distributed in China, Pakistan, Bangladesh, Malaysia, etc.

## 生长习性

喜光，喜温暖，耐旱，不耐寒霜；适生温度25~30℃。在年降水量700~2000mm的地区生长良好。适生于微酸性的壤土或砂壤土。

## 用途、价值

为热带常见水果，食用价值高，富含多种营养物质。树形优美，常用于园林绿化。果皮可入药，叶和树皮可作染料；木材坚硬，耐海水，适合作舟车或家具等。

## GROWTH HABIT

Light-favored and thermophilus, unhardy, drought resistance; suitable for 25-30°C. It grows well in areas with annual rainfall of 700-2000mm. Suitable for slightly acidic loam or sandy loam.

## APPLICATION,VALUE

It is a common tropical fruit with high edible value and rich in various nutrients. Beautiful tree, often used for landscaping. Peel can be used as medicine, and leaves and bark can be used as dyes; wood is hard and resistant to seawater, and is suitable for boats, cars and furniture.

# 39 / 悬铃木

## 植物形态

落叶大乔木。树高5~30m。树皮不规则片状剥落，剥落后呈灰绿色，光滑。单叶互生，叶片阔卵形，3~5掌状分裂，边缘有不规则尖齿和波状齿，基部近三角状心形，长7~9cm，宽4~6cm，嫩时有星状毛。头状花序球形，无柄，基部有长茸毛；花期4~5月。圆球形头状果序，宿存花柱突出呈刺状，小坚果之间有黄色茸毛，突出头状果序外；果期9~10月。

## 生态分布

原产于欧洲。分布于中国、巴基斯坦、印度等地。

## 生长习性

喜光，喜湿润温暖气候，较耐寒；适生于微酸性或中性、排水良好的土壤。

## 用途、价值

园林绿化树种。适应性强，耐修剪整形，对多种有毒气体抗性较强，是优良的庭荫树、行道树。

## PLANT MORPHOLOGY

Large deciduous tree. Plant height 5-30m. bark peeled off irregularly, gray green and smooth. Single leaves alternate, broad ovate leaves, 3-5 palmate divisions, irregular tines and wavy teeth at the edges, nearly triangular heart-shaped at the base, 7-9cm in length, 4-6cm in width, stellate when tender hair. flower head spherical, sessile, with long hairs at the base; flowering period from April to May. The spherical capitate infructescence, the persistent style prominent and thorn-like, with yellow hairs between the nutlets, protruding outside the capitate infructescence; fruiting period from September to October.

## ECOLOGICAL DISTRIBUTION

Native to Europe. Distributed in China, Pakistan, India, etc.

## GROWTH HABIT

It is suitable for slightly acidic or neutral soil with good drainage.

## APPLICATION, VALUE

It is a garden tree species. It is a good shade tree and street tree with strong adaptability, pruning and shaping resistance and strong resistance to various toxic gases.

# 40 / 构树

*Broussonetia papyrifera*
桑科 Moraceae　构属 *Broussonetia*
别名：构桃树、沙纸树、谷木、谷浆树、假杨梅

## 植物形态

落叶乔木。树高1.5~20m。全株含乳汁。树皮暗灰色；小枝密被灰色粗毛。叶螺旋状排列，广卵形至长椭圆状卵形，长6~18cm，宽5~9cm；基部具粗锯齿，不裂或2~5裂；三出脉，叶柄被糙毛。花雌雄异株，雄花柔荑花序，苞片披针形，被毛，长3~8cm；雌花头状花序，苞片棍棒状，顶端被毛，花期4~5月。聚花果球形，熟时橙红色，肉质；果期6~7月。

## 生态分布

原产东亚。分布于中国、越南、巴

## PLANT MORPHOLOGY

Deciduous tree. Plant height 1.5-20m. whole plant containing milk. Bark dark gray; branchlets densely covered with gray shag. leaves arranged spirally, broadly ovate to oblong-ovate, 6-18cm in length and 5-9cm in width; base coarsely serrated, not split or 2-5 split; three veins, petiole with coarse hairs. flowers e dioecious, male catkins, bracts lanceolate, hairy, 3-8cm in length; female flower heads, bracts club-like, apex hairy, flowering period from April to May. fruit spherical, orange-red when ripe, fleshy, fruiting period from June to July.

## ECOLOGICAL DISTRIBUTION

Native to East Asia. Distributed in China, Vietnam, Pakistan, Japan, India, etc.

基斯坦、日本、印度等地。

## 生长习性

喜光，耐干旱瘠薄，适应性强。抗逆性强，耐烟尘，抗大气污染能力强。多生于石灰岩山地，也能在酸性土、中性土及水边生长。

## 用途、价值

可作行道树。能抗多种有毒气体，也可作荒滩、偏僻地带及污染严重的工厂区绿化树种。叶营养成分十分丰富，加工后可用于生产饲料。可药用，有补肾、利尿、强筋骨等功效。

## GROWTH HABIT

Favoring sunshine, resistant to drought and barren, and has strong adaptability. Strong resistance, smoke and dust resistance, strong ability to resist air pollution. It is mainly grown in limestone mountains, and can also grow in acid soil, neutral soil and water edge.

## APPLICATION,VALUE

It can be used as street tree. It can resist many kinds of toxic gases and can be used as greening tree species in barren beaches, remote areas and factories with serious pollution. The leaves are rich in nutrients and can be used to produce feed after processing. It can be used for medicine. It has the functions of tonifying kidney, diuresis and strengthening muscles and bones.

# 41 / 愈疮木

*Guaiacum officinal*
蒺藜科 Zygophyllaceae　愈疮木属 *Guaiacum*
别名：铁梨木、圣檀木、铁木

## 植物形态

常绿阔叶乔木。树高2~12m。树干常弯曲，树皮淡褐色并具灰绿色斑纹。偶数羽状复叶，对生，长6~13cm；叶革质，小叶2~3对，倒卵形或椭圆形，长2~4cm，宽2cm；托叶小。花量大，数朵簇生于叶腋或近顶生；开放时蓝紫色，之后逐渐转为浅蓝色或白色，花梗细，长3~4cm，花期12月至翌年9月。倒心形蒴果，两侧扁平，幼时红色，熟时橘黄色。种子肾形，种皮红色，肉质；1枚。

## PLANT MORPHOLOGY

Evergreen broad-leaved tree. Plant height 2-12m. trunk often curved, and bark light brown with gray-green markings. Even-pinnate compound leaves, opposite, 6-13cm in length; leaf leathery, 2-3 pairs of lobules, obovate or elliptical, 2-4cm in length, 2cm in width; stipules small. Flowers large, with several clusters growing in the leaf axils or near the terminal; blue-purple when opening, and then gradually turning to light blue or white. peduncle thin, 3-4cm in length, and flowering period from December to September of the following year. Inverted heart-shaped capsule, flat on both sides, red when young, orange when ripe. Seeds kidney-shaped, red seed coat, fleshy; 1 piece.

## 生态分布

原产于热带美洲。分布于中国、巴基斯坦、墨西哥、委内瑞拉等地。

## 生长习性

喜光树种，耐热、耐旱、耐湿、耐盐碱、耐瘠薄，不耐水渍，不耐寒，适生于高温湿润、水热平衡的气候。适生温度20~32℃，深根性树种，根系发达，对土壤要求不严。

## 用途、价值

常用园林绿化观赏树种，可作园景树、行道树、庭院树。具有防风固沙、保持水土的作用。木材坚硬，是硬重木材用材树种，是水下工程设施材料，船舶工业特种用材；可作轴承、滑轮、家具、工艺品等。可药用，可用于治疗痛风、风湿病、扁桃体炎、咳嗽、皮肤病等症。花和叶可制茶。是牙买加的国花。

## ECOLOGICAL DISTRIBUTION

Native to tropical America. Distributed in China, Pakistan, Mexico, Venezuela, etc.

## GROWTH HABIT

It is suitable for the climate of high temperature and humidity balance. The suitable temperature for growth is 20-32°C, deep rooted tree species, well-developed roots, and no strict requirements for soil.

## APPLICATION,VALUE

It is commonly used as ornamental trees for landscaping, and can be used as garden trees, street trees, and garden trees. It has the function of preventing wind and sand fixing and maintaining water and soil. The wood is hard, as a kind of hard and heavy wood, a material for underwater engineering facilities and a special material for the shipbuilding industry; it can be used as bearings, pulleys, furniture, handicrafts, etc. It is medicinal and can be used to treat gout, rheumatism, tonsillitis, cough, skin diseases and other diseases. Flowers and leaves can make tea. It is the national flower of Jamaica.

# 42 / 糖胶树

*Alstonia scholaris*

夹竹桃科 Apocynaceae　鸡骨常山属 *Alstonia*

别名：面条树、大枯树、鹰爪木、灯架树、黑板树

## 植物形态

　　常绿乔木。树高5~20m。枝轮生，具乳汁，无毛。叶轮生，倒卵状长圆形、倒披针形或匙形，长7~28cm，宽2~11cm。聚伞花序，顶生，花白色，被柔毛；总花梗长4~7cm；花冠高脚碟状，中部以上膨大，内面被柔毛，柱头顶端2深裂；花期6~11月。蓇葖果，外果皮近革质，灰白色，长20~57cm，直径0.2~0.5cm。种子长圆形，红棕色，两端被红棕色长缘毛；果期10月至翌年4月。

## 生态分布

　　原产斯里兰卡和澳大利亚。分布于

## PLANT MORPHOLOGY

　　Evergreen tree. Plant height 5-20m. Branches whorled, milky, glabrous. Leaves whorled, obovate oblong, oblanceolate or spatulate, 7-28cm in length and 2-11cm in width. Cymes, terminal, white pilose flowers; total pedicel 4-7cm in length; corolla high foot saucer shaped, inflated above middle, inner surface pilose, stigma top 2-lobed; flowering period from June to November. Follicles, exocarp nearly leathery, grayish white, 20-57cm in length, 0.2-0.5cm in diameter; seeds oblong, reddish brown, with long reddish brown hairs at both ends, fruiting period from October to April of the following year.

## ECOLOGICAL DISTRIBUTION

　　Native to Sri Lanka and Australia. Distributed in China, Nepal, India, Malaysia, Indonesia,

中国、尼泊尔、印度、马来西亚、印度尼西亚、菲律宾、巴基斯坦等地。

## 生长习性

喜湿润、耐阴、耐强光，对土壤适应性强。

## 用途、价值

树形美观，观赏价值高，常用作行道树、庭荫树等。乳汁干后为胶，可制口香糖。树根、皮及叶均可入药，具有镇咳、消炎退热功效，治气管炎、哮喘等症。汁液含生物碱，具轻微毒性。

Philippines, Pakistan, etc.

**GROWTH HABIT**

Favoring moist, tolerant to shade, tolerant to strong light, and has strong adaptability to soil.

**APPLICATION,VALUE**

It has beautiful shape and high ornamental value. It is often used as street tree and court shade tree. Dry milk for gum. Tree root, bark and leaf can be used as medicine, with antitussive, antiphlogistic and antipyretic, treating tracheitis and asthma. The juice contains alkaloids with slight toxicity.

## 43 / 侧柏

*Platycladus orientalis*
柏科 Cupressaceae　侧柏属 *Platycladus*
别名：香柯树、香树、扁桧、香柏、黄柏

### 植物形态

常绿乔木。树高1~20m。幼树树冠尖塔形，老树树冠广圆形。树皮浅灰褐色。叶鳞形，交互对生，背面有腺点，小枝细且扁平。球花单生枝顶，雌雄同株，雄球花黄色、卵圆形，雌球花近球形，蓝绿色，被白粉；花期3~4月。球果卵状椭圆形，成熟时褐色；种鳞木质，背部顶端下方有一弯曲的钩状尖头。种子椭圆形或卵圆形，长0.4~0.6cm，灰褐或紫褐色，顶端或有短膜，种脐大而明显，球果10月成熟。

### 生态分布

原产中国。分布于朝鲜、巴基斯坦等地。

### 生长习性

喜光，耐干旱瘠薄，耐高温；浅根性，抗风能力较弱，适应性强，对土壤要求不严；萌芽力强。

### 用途、价值

常作园林绿化树种。有吸附尘埃、净化空气的作用。木材富含树脂，耐腐性强，可供建筑、家具、文具等用材。根皮、种子、果实、叶均可入药，具有治肾热病、咳嗽、胃出血等功效。

### PLANT MORPHOLOGY

Evergreen tree. Plant height 1-20m. e Crown of young trees steeple-shaped, and crown of old trees wide round. Bark grayish brown. Leaves scale-shaped, alternately opposite, with glandular dots on the back, and small and flat branchlets. Top of the cone solitary, monoecious, the male cone yellow, ovoid, female cone sub globose, blue-green, and white powder; the flowering period from March-April. Cones ovate-elliptic, brown when mature; seed scales woody, with a curved hook-shaped tip under the top of the back. Seeds oval or ovoid, 0.4-0.6cm in length, grayish brown or purple-brown, with short membranes on the top, large and obvious hilum, and ripe autumn fruits in October.

### ECOLOGICAL DISTRIBUTION

Native to China. Distributed in North Korea, Pakistan, etc.

### GROWTH HABIT

Favoring sunshine, drought and barren, high temperature and shallow root, weak wind resistance, strong adaptability, not strict requirements on soil, and strong germination ability.

### APPLICATION,VALUE

It is a commonly used landscaping tree species. It can absorb dust and purify the air. The wood is rich in resin and has strong corrosion resistance, which can be used for construction, furniture, stationery and other materials. Root bark, seeds, fruits and leaves can all be used as medicine. It has the effects of curing kidney fever, coughing, and stomach bleeding.

# 44 / 石榴

*Punica granatum*

千屈菜科 Lythraceae  石榴属 *Punica*

别名：若榴木、花石榴

## 植物形态

落叶乔木。树高3~5m。树干灰褐色，上有瘤状突起。枝顶常成尖锐长刺，幼枝具棱角。叶对生，长圆状披针形，长2~9cm，宽1~2cm。花钟状或筒状，多为红色，萼片硬，肉质，管状；花瓣倒卵形，花有单瓣、重瓣之分，宿存；花期5~10月。果近球形，直径5~15cm，淡黄褐或淡黄绿色，有时白色，暗紫色。种子多数，钝角形，肉质外种皮淡红色至乳白色，直径0.2~0.6cm。

## PLANT MORPHOLOGY

Deciduous tree. Plant height 3-5m. trunk grayish brown with tuberculate protuberances. Top of the branch often sharp and long, and young branch angular. Leaves opposite, oblong lanceolate, 2-9cm in length and 1-2cm in width. Flowers bell shaped or tubular, mostly red, hard sepals, fleshy, tubular; petals obovate, single or double, persistent, flowering period from May to October. Fruit globose likely, 5-15cm in diameter, yellowish brown or yellowish green, sometimes white, dark purple. Seeds numerous, obtuse angle, fleshy exocarp reddish to milky white, 0.2-0.6cm in diameter.

## 生态分布

原产巴尔干半岛至伊朗及其邻近地区。分布于巴基斯坦、中国等地。

## 生长习性

喜温暖向阳的环境，耐旱、耐寒、耐瘠薄，不耐涝和阴，适生于排水良好的砂壤土。

## 用途、价值

常见果树及观赏树种。可大田种植，也可供公园、家庭、行道种植美化环境。药食同源，营养丰富，维生素含量高，果实可榨汁，含丰富的矿物质、花青素，具有补水、保护眼睛等作用；也有驱虫、抗菌的功效，可治腹泻、中耳炎、扁桃体炎、月经不调等症。树皮、根皮和果皮可提制栲胶。根皮含有生物碱，具毒。

## ECOLOGICAL DISTRIBUTION

Native to the Balkan Peninsula to Iran and its adjacent areas. Distributed in Pakistan, China, etc.

## GROWTH HABIT

Favoring warm and sunny environment, is resistant to drought, cold, barrenness, waterlogging and shade, and is suitable for well-drained sandy soil.

## APPLICATION,VALUE

It is a common fruit tree and ornamental tree species. But large field planting, and can be used for park, courtyard and sidewalk planting to beautify the environment. Medicine food homology. Rich in nutrition, high in vitamin content, the fruit can be squeezed, rich in minerals, anthocyanins, and has the effects of replenishing water and protecting eyes. It has deworming and antibacterial effects, and can cure diarrhea, otitis media, tonsillitis and irregular menstruation. The bark, root bark and peel can be extracted from tannin extract. Root bark contains alkaloids, which is toxic.

## 45 / 南洋杉

*Araucaria cunninghamii*
南洋杉科 Araucariaceae　南洋杉属 *Araucaria*
别名：猴子杉、肯氏南洋杉、细叶南洋杉

### 植物形态

常绿乔木。树高3~30m。树皮灰褐色或暗灰色，粗糙横裂；幼树冠尖塔形，老则成平顶状，侧生小枝密生，下垂，近羽状排列。叶二型，幼树和侧枝的叶排列疏松开展，钻状、针状、镰状或三角状，大树及花果枝上的叶排列紧密而叠盖，斜上伸展，微向上弯，卵形、三角状卵形。雄球花单生枝顶。球果卵形或椭圆形，长6~10cm，径4~8cm。种子椭圆形，两侧具膜质翅。

### 生态分布

原产南美洲、大洋洲及太平洋群岛、大洋洲昆士兰等东南沿海地区。分布于中国、巴基斯坦等地。

### 生长习性

喜温暖湿润、光照充足的环境，不耐寒，忌干旱，适生于气温25~30℃、相对湿度70%以上的环境；要求疏松肥沃、腐殖质含量较高、排水透气性强的土壤。

### 用途、价值

园林绿化观赏树种。树形优美，是世界著名的庭院树之一，常用于室内盆栽装饰。也是重要的用材树种，材质优良，可用于建筑、器具、家具等。

### PLANT MORPHOLOGY

Evergreen tree. Plant height 3-30m. Bark grayish brown or dark gray, rough and transversely split; crown of young trees tower-shaped, old trees flat top-shaped, lateral branchlets densely distributed, drooping, and nearly pinnate. Leaves of young trees and lateral branches loosely arranged and spread, drill like, needle like, sickle shaped or triangular in shape. Leaves on big trees and flower and fruit branches arranged closely and overlapped, extending obliquely, slightly upward, ovate, triangular ovate. Male cone having a single shoot apex. Cones ovate or elliptic, 6-10cm in length and 4-8cm in diameter; seeds elliptic, with membranous wings on both sides.

### ECOLOGICAL DISTRIBUTION

Native to South America, Oceania, the Pacific Islands and other southeast coastal areas. Distributed in China, Pakistan, etc.

### GROWTH HABIT

Favoring warm and humid environment with sufficient sunlight, unhardy and avoiding drought, suitable for 25-30°C temperature and 70% relative humidity or more; requires loose and fertile, high humus content, and strong drainage and air permeability.

### APPLICATION,VALUE

Ornamental tree species for landscaping. One of the most famous garden trees in the world. Indoor potted plant decoration. an important timber tree species with excellent material, applicable in buildings, appliances, furniture, etc.

# 46 / 棟

*Melia azedarach*
棟科 Meliaceae　棟属 *Melia*
别名：苦棟树、森树、紫花树、棟树

## 植物形态

落叶或常绿乔木。树高5~20m。树皮灰褐色。二至三回奇数羽状复叶，叶互生；小叶对生，卵形或披针形，锯齿粗钝。花两性，腋生圆锥花序，花淡紫色，倒卵状匙形，两面均被毛；花期4~5月。核果球形至椭圆形，长1~2cm，宽5~20mm，熟时黄色，果期10~12月。种子椭圆形，黑色。

## 生态分布

原产中国。分布于巴基斯坦、印度、越南等地。

## PLANT MORPHOLOGY

Deciduous or evergreen tree. Plant height 5-20m. Bark taupe. 2-3 times of odd-numbered pinnate compound leaves, alternate leaves; leaflets opposite, ovate or lanceolate, serrate and blunt. flowers bisexual, axillary panicles, lavender flowers, obovate spoon-shaped, hairy on both sides; flowering period April-May. drupe spherical to oval, 1-2cm in length, 5-20mm in width, yellow when ripe, and fruiting from October to December. Seeds oval, black.

## ECOLOGICAL DISTRIBUTION

Native to China. Distributed in Pakistan, India, Vietnam, etc.

## 生长习性

喜光，喜温暖，较耐寒。不耐阴，不耐旱，忌涝；对土壤要求不严。

## 用途、价值

作行道树，是良好的城市及矿区绿化树种；能耐烟尘及杀菌，能大量吸收有毒气体，具有空气净化作用。与其他树种混栽，能起到对树木虫害的防治作用；木材可作家具、建筑、乐器等；鲜叶可作农药；根皮粉调醋可治疥癣；果核仁油可供制油漆、润滑油和肥皂。

## GROWTH HABIT

Light-favored, thermophilic and hardy. Not tolerant of shade, drought and waterlogging, not strict to the soil quality.

## APPLICATION,VALUE

As a street tree, good tree species for greening in cities and mining areas; resistant to smoke, dust and sterilization, absorb a large amount of toxic gases, air purification. Mixed planting with other tree species can prevent tree pests; wood applicable in furniture, construction, musical instruments, etc.; fresh leaves applicable in pesticides; root bark powder applicable in vinegar to cure mange; kernel oil applicable in paint, lubricating oil and soap.

# 47 / 沙枣

*Elaeagnus angustifolia*
胡颓子科 Elaeagnaceae　胡颓子属 *Elaeagnus*
别名：银柳胡颓子、银柳、桂香柳、刺柳、香柳、七里香

## 植物形态

落叶乔木。树高5~10m。有时具枝刺，刺长30~40mm，棕红色，发亮；幼枝、嫩叶、花及果实均密被银白色鳞片。叶薄纸质，披针形，叶背灰白色，密被白色鳞片。花银白色，常簇生于叶腋；萼筒钟形，内面被白色星状柔毛；花盘明显，圆锥形，包围花柱的基部；花期5~6月。果椭圆形，粉红色；果肉乳白色；长9~12mm，直径6~10mm；果期9月。

## 生态分布

原产中国。分布于巴基斯坦、印

## PLANT MORPHOLOGY

Deciduous tree. Plant height 5-10m. Sometimes with branch spines, 30-40mm in length, brownish red, shiny; young branches, young leaves, flowers and fruits are densely covered with silver white scales. Leaves thin papery, lanceolate, gray white on the back of leaves, densely covered with white scales. Flowers silvery white, often clustered in leaf axils; calyx tube campanulate, inner surface covered with white stellate pilose; disk distinct, conical, surrounding the base of style; flowering period from May to June. Fruit oval and pink; flesh milky white; 9-12mm in length, 6-10mm in diameter; fruiting period September.

## ECOLOGICAL DISTRIBUTION

Native to China. Distributed in Pakistan, India, Vietnam, etc.

度、越南等地。

## 生长习性

适应力强，山地、平原、沙滩、荒漠均能生长；对土壤、气温、湿度要求较低。

## 用途、价值

根蘖性强，能保持水土，抗风沙，防止干旱，调节气候，改良土壤，常用来营造防护林、防沙林、用材林和风景林。果肉可食；果实和叶可作饲料，花可提精油和酿蜜，对荒漠地区农业稳产丰收具有很大作用。木材可作家具、农具，亦可作燃料，是沙漠地区燃料的主要来源之一；果实、叶、根可入药，具有治痔疮、治肺炎、气短等功效。

## GROWTH HABIT

Strong adaptability, growing in mountains, plains, beaches and deserts; not strict to the soil quality, temperature and humidity.

## APPLICATION.VALUE

Strong root tillers, maintain water and soil, resist wind and sand, prevent drought, regulate climate, and improve soil. build shelter forests, sand prevention forests, timber forests and scenic forests. Pulp edible; fruits and leaves fodder, flowers extract essential oil and brew honey, applicable in the stable and good harvest of agriculture in desert areas. Wood applicable in in furniture, agricultural tools, and fuel. One of the main sources of fuel in desert areas. Fruits, leaves, and roots applicable in medicine, and effective in curing hemorrhoids, pneumonia, and shortness of breath.

# 48 / 麻风树

*Jatropha curcas*

大戟科 Euphorbiaceae　麻风树属 *Jatropha*
别名：黄肿树，假白榄

## 植物形态

小乔木，高2~5m，具水状液汁，树皮平滑；枝条苍灰色，无毛，疏生突起皮孔，髓部大。叶纸质，近圆形至卵圆形，长7~18cm，宽6~16cm，顶端短尖，基部心形，全缘或3~5浅裂，上面亮绿色，无毛，下面灰绿色，初沿脉被微柔毛，后变无毛；叶柄长6~18cm；托叶小。花序腋生，苞片披针形；花瓣长圆形，黄绿色，合生至中部，内面被毛。蒴果椭圆状或球形，长2.5~3cm，黄色。种子椭圆状，长1.5~2cm，黑色。花期9~10月。

## PLANT MORPHOLOGY

Small tree. Plant height 2-5m, with watery sap, smooth bark; branches pale gray, glabrous, sparsely protruding lenticels, large pith; leaves papery, nearly round to ovoid, 7-18cm long, 6-16cm wide, short pointed tip, heart-shaped at base, whole or 3-5 lobed, bright green on top, glabrous, gray-green below, puberulent along veins at first, then glabrous; petiole 6-18cm long; stipules small. Inflorescences axillary, bracts lanceolate; petals oblong, yellow-green, connate to the middle, and the inner surface hairy. Capsules oval or spherical, 2.5-3cm long, yellow; seeds elliptic, 1.5-2cm long, black. Flowering period from September to October.

## ECOLOGICAL DISTRIBUTION

Native to tropical America. Now widely

## 生态分布

原产美洲热带；现广布于全球热带地区。

## 生长习性

喜光植物，根系粗壮发达，具有很强的耐干旱耐瘠薄能力，对土壤条件要求不严，生长迅速，抗病虫害，适宜中国秦岭淮河以南地区种植。

## 用途、价值

种子含油量高，油供工业或医药用。

distributed in tropical regions of the world.

### GROWTH HABIT

It is a light-loving plant with strong and well-developed root system, strong drought and barren tolerance, lax requirements for soil conditions, rapid growth, resistance to pests and diseases, suitable for planting in areas south of the Qinling Mountain-Huaihe River, China.

### APPLICATION, VALUE

Its seeds have high oil content, and the oil is used for industry or medicine.

# 49 / 女贞

*Ligustrum lucidum*
木樨科 Oleaceae　女贞属 *Ligustrum*
别名：白蜡树、女桢、将军树

## 植物形态

常绿乔木。高5~20m。树皮灰褐色，枝条具皮孔。叶片革质，对生，卵形或椭圆形，长6~17cm，宽3~8cm。圆锥花序，顶生；花序轴紫色或黄棕色；花序基部苞片常与叶同型，凋落；花无梗或近无梗；花期5~7月。核果肾形或近肾形，长7~10mm，直径4~6mm，深蓝黑色，成熟时呈红黑色，被白粉；果期7月至翌年5月。

## 生态分布

分布于中国、朝鲜、印度、尼泊尔、巴基斯坦等地。

## PLANT MORPHOLOGY

Evergreen tree. Plant height 5-20m. Bark grayish brown, branches with lenticels. Leaves leathery, opposite, ovate or elliptic, 6-17cm in length and 3-8cm in width. Panicle, terminal; inflorescence axis purple or yellowish brown; basal bracts of inflorescence often the same type as leaves, withered; flowers sessile or nearly sessile; flowering period from May to July. Drupe reniform or subreniform, 7-10mm in length 4-6mm in diameter, dark blue black, red and black at maturity, with white powder. Fruiting period from July to May of the following year.

## ECOLOGICAL DISTRIBUTION

Distributed in China, North Korea, India, Nepal, Pakistan, etc.

## 生长习性

喜温暖湿润气候，喜光耐阴，耐寒。适宜在湿润、背风、向阳的地方栽种，适生于砂质壤土。

## 用途、价值

可作行道树，种植在庭院，或用作绿篱，是园林中常用的观赏树种。对大气污染的抗性较强，对二氧化硫、氯气、氟化氢及铅蒸气均有较强抗性，能忍受较高的粉尘、烟尘污染。果肉可供酿酒或制酱油；种子油可制肥皂；花可提取芳香油。

## GROWTH HABIT

Favoring warm and humid climate, light, shade and cold. It is suitable for planting in humid, leeward and sunny places, and suitable for growing in sandy loam.

## APPLICATION,VALUE

It can be used as street tree, planted in the courtyard, used as hedge, is a common ornamental tree species in the garden. It has strong resistance to air pollution, strong resistance to sulfur dioxide, chlorine gas, hydrogen fluoride and lead vapor, can tolerate high dust and smoke pollution. The pulp can be used to make wine or soy sauce; the seed oil can be used to make soap; the flower can be used to extract aromatic oil.

# 50 / 垂枝红千层

桃金娘科 Myrtaceae　红千层属 Callistemon

*Callistemon viminalis*

别名：串钱柳

## 植物形态

常绿小乔木。高1.5~6m。树皮暗灰色。叶互生，条形，长3~8cm，宽2~5mm，坚硬，无毛，有透明腺点，中脉明显，无柄。穗状花序，红色，无梗；萼筒钟形，裂片5；花瓣5，脱落；雄蕊多数，红色；花期3~10月。蒴果顶端开裂，半球形，直径1cm；果熟期8~12月。

## 生态分布

原产澳大利亚。分布于中国、巴基斯坦等地。

## PLANT MORPHOLOGY

Evergreen small tree. Plant height 1.5-6m. bark dark gray. Leaves alternate, strip, 3-8cm in length, 2-5mm in width, hard, glabrous, with hyaline glandular spots, obvious midvein, sessile. Inflorescence spikes, red, sessile; calyx tube bell shaped, lobes 5; petals 5, abscission; stamens numerous, red; flowering period from March to October. Capsule hemispherical and 1cm in diameter. Fruiting period from August to December.

## ECOLOGICAL DISTRIBUTION

Native to Australia. distributed in China, Pakistan, etc.

## GROWTH HABIT

Light-favored, favoring warm and humid

## 生长习性

喜光，喜温暖湿润气候，耐阳光暴晒，耐-5℃低温和45℃高温，适生温度在25℃左右。适生于肥沃、酸性土壤。

## 用途、价值

是园林绿化观赏植物，花形美丽奇特，可作切花或大型盆栽；是庭院美化树、行道树、风景树；还可作防风林。极耐旱、耐瘠薄，可在荒山或公园等处栽培，用于生态景观建设。

climate, sunlight resistance, low temperature below - 5℃ and high temperature of 45℃, suitable for 25℃ temperature and growing in fertile and acid soil.

### APPLICATION,VALUE

Ornamental plant for landscaping, with beautiful and unique flowers, used as cut flowers or large-scale potted plants; garden beautification tree, street tree, landscape tree; used as a windbreak forest. cultivated in barren mountains or parks for ecological landscape construction.

# 51 / 黄钟树

*Tecoma stans*

紫葳科 Bignoniaceae　黄钟花属 *Tecoma*
别名：黄钟花

## 植物形态

常绿小乔木。高4~10m。树皮淡褐色。叶对生，奇数羽状复叶；小叶3~7枚，椭圆状披针形，长7~10cm，宽1~3cm，边缘有粗锯齿。总状花序，顶生；花萼合生成杯状；花冠黄色或橙黄色，喇叭状，先端5裂，裂片向外反卷；花期4~9月。蒴果长柱形，长10~20cm，表面有稀疏的小突点，熟时变褐色；果期秋季。种子扁平椭圆形，长6mm，每端有长6mm的翅。

## 生态分布

原产南美洲和西印度群岛、阿根廷北部。分布于中国、巴西、巴基斯坦等地。

## 生长习性

喜高温，耐盐碱。适生温度22~30℃；不耐寒，根部能耐-20℃的低温。喜光，耐干旱。对土壤要求不严，适生于肥沃而排水良好的土壤。

## 用途、价值

是园林绿化观赏树种。花色鲜艳，花期较长，可作花园、公园、庭院的美化栽培；可作盆栽观赏；耐修剪，可剪成各形状或作花墙。根可入药，作利尿、驱虫和兴奋刺激剂。

## PLANT MORPHOLOGY

Evergreen small tree. Plant height 4-10m. bark light brown. Leaves opposite, odd pinnate compound; leaflets 3-7, elliptic lanceolate, 7-10cm in length, 1-3cm in width, margin coarsely serrate. Inflorescence racemose, terminal; calyx united into cup; corolla yellow or orange yellow, trumpet shaped, apex 5-lobed, lobes rolled outward, flowering period from April to September. capsule long columnar, 10-20cm in length, with sparse small protuberances on the surface, turning brown at maturity. fruiting period autumn. Seeds flat oval, 6mm in length, with wings 6mm in length at each end.

## ECOLOGICAL DISTRIBUTION

Native to South America, West Indies and northern Argentina. Distributed in China, Brazil, Pakistan, etc.

## GROWTH HABIT

Thermophilic, salt-alkali resistance. suitable for 22-30°C temperature; unhardy, root tolerate for -20°C temperature. Light-favored, drought resistance. suitable for fertile and well drained soil.

## APPLICATION,VALUE

Ornamental tree species for landscaping. Flower bright in color, long flowering period, used to beautify gardens, parks and courtyards. And used as potted ornamental plants. Can be cut into various shapes or used as flower walls. Root applicable in medicine, diuretic, anthelmintic and stimulant.

第二章

# 灌木

PART 2　SHRUB

# 01 / 三角梅

*Bougainvillea spectabilis*
别名：叶子花

## 植物形态

灌木。少数种可长成15m以上的高大乔木。枝叶密生柔毛；刺腋生，下弯。叶片椭圆形或卵形。花序腋生或顶生；苞片形似花瓣，椭圆状卵形，暗红色或淡紫红色；花被管狭筒形，绿色，密被柔毛；裂片开展，黄色，长3~5mm。果实长1~2cm，密生毛。

## 生态分布

原产热带美洲。分布于中国、巴基斯坦、印度、巴西等地。

## PLANT MORPHOLOGY

Shrubs. A few species growing into tall trees more than 15m. Branches and leaves densely pilose; thorn axillary, curved down. leaves oval or ovoid. Inflorescence axillary or terminal; bracts like petals, elliptic ovoid, dark red or lavender; perianth tube narrow and tube-shaped, green, densely pilose; lobes open, yellow, 3-5mm in length. Fruit 1-2cm in length and densely hairy.

## ECOLOGICAL DISTRIBUTION

Native to tropical America. Distributed in China, Pakistan, India, Brazil, etc.

## GROWTH HABIT

Warm-favored, wet-favored, light-favored. Unhardy, barren, drought and saline-alkali

## 生长习性

喜温、喜湿、喜光。不耐寒，耐瘠薄，耐干旱，耐盐碱；对土壤要求不严，在肥沃、疏松、排水好的砂质壤土生长旺盛。

## 用途、价值

是一种环保绿化植物。常用于庭院绿化，作花篱、棚架植物，花坛、花带的配置，也可作切花。花可入药，具有清热解毒、调和气血的功效。

resistance; it does not have strict requirements on the soil, and grows vigorously in fertile, loose, and well-drained sandy loam.

### APPLICATION,VALUE

Environmentally friendly green plant. Often used for garden greening, making flower hedges, trellis plants, flower beds, flower belts, and cutting flowers. Flowers applicable in medicine, effective in detoxification, clearing away heat, and reconciling blood.

# 02 / 树牵牛

*Ipomoea carnea* subsp. *fistulosa*
旋花科 Convolvulaceae　虎掌藤属 *Ipomoea*
别名：南美旋花

## 植物形态

散状灌木。树高1~3m。小枝圆柱形或有棱，密被微柔毛，有白色乳汁。叶宽卵形或卵状长圆形，叶长6~25cm，宽4~17cm，具小短尖头。聚伞花序；花梗被微柔毛；花冠漏斗状，淡红色，内面至基部深紫色，花冠管基部缢缩，花冠管被微柔毛。果实为蒴果，卵形或球形。种子4枚或较少，表面被绢状长柔毛。

## 生态分布

原产美洲。分布于中国、巴基斯坦、墨西哥、巴西等地。

## PLANT MORPHOLOGY

Scattered shrub. Plant height 1-3m. Branches cylindrical or ribbed, densely puberulent, with white milk. Leaves broadly ovate or ovoid, 6-25cm in length and 4-17cm in width, with short pointed tips. Inflorescences cymes; peduncle puberulent; Corolla funnel-shaped, light red, dark purple from the inner surface to the base. Base of the corolla tube constricted, corolla tube puberulent. Capsule, ovoid or globose. Seeds 4 or less, surface sericeous villous.

## ECOLOGICAL DISTRIBUTION

Native to America. Distributed in China, Pakistan, Mexico, Brazil, etc.

## 生长习性

喜光，适生温度20~35℃，耐旱、耐瘠薄；对土壤要求不严，适生于疏松、排水良好的砂质壤土。

## 用途、价值

易栽培，花期长，可作盆栽，是园林绿化的优良树种。

## GROWTH HABIT

Light-favored, suitable for 20-35℃ temperature, barren and drought resistance; not strict to the soil quality, suitable for loose sandy loam with good drainage.

## APPLICATION, VALUE

Easy to cultivate, long flowering period, used as a potted plant. Excellent tree species for landscaping.

# 03 / 牛角瓜

夹竹桃科 Apocynaceae
*Calotropis gigantea*
牛角瓜属 *Calotropis*
别名：五狗卧花

## 植物形态

常绿灌木，高达5m。幼枝被灰白色茸毛。叶倒卵状长圆形或长圆形，长5~30cm，先端钝，基部心形，两面被灰白色茸毛，老时渐脱落；侧脉4~8对，疏离；叶柄长0.1~0.5cm。聚伞花序伞状，腋生和顶生；花序梗和花梗被灰白色茸毛，花梗长2~3cm；花萼裂片卵圆形；花冠紫蓝色，辐状，直径3~4cm；裂片卵圆形，长1~2cm，宽1cm，急尖；副花冠裂片比合蕊柱短，顶端内向，基部有距。种子广卵形，顶端具白色绢质种毛；蓇葖单生，膨胀，端部外弯，长5~10cm，直径

## PLANT MORPHOLOGY

Evergreen shrub. Up to 5m high. Young branches grayish white villi. Leaves obovate oval or oblong, 5-30cm long, blunt apex, heart-shaped at the base, grayish white villi on both sides, gradually falling off, lateral veins 4-8 pairs, detached; petiole 1-5mm long. Cymes umbrella shape, axillary and terminal; peduncle and pedicel gray-white tomentose, pedicel 2-3cm long; calyx lobes oval; corolla purple-blue, radial, 3-4cm in diameter; lobes oval, 1-2cm long, 1cm wide, acutely pointed; corona lobes shorter than stamens, the apex is inward, the base is spaced. Seeds broadly ovate, with white silky seed hairs at the top; follicles solitary, swollen, with curved ends, 5-10cm long, 3cm in diameter, pubescent. Flowering and fruiting periods almost all year round.

3cm，被短柔毛。花果期几乎全年。

## 生态分布

分布于中国、印度、巴基斯坦、斯里兰卡、缅甸、越南和马来西亚等地。

## 生长习性

喜光植物，适宜生长在20~35℃的温度下；生长于低海拔向阳山坡、旷野地及海边。

## 用途、价值

全株可作绿肥。乳汁干燥后可用作树胶原料，还可制鞣料及黄色染料。根、茎、叶和果等均可药用，具有消炎、抗菌、化痰和解毒等作用。茎叶的乳汁有毒，含有多种强心苷，供药用；树皮可治癫痫。茎皮纤维可用于制绳索、造纸、人造棉和织麻布、麻袋等，其种毛可作填充物及丝绒原料；纤维用于纺织生产可替代棉纤维。果形奇特，有一定的观赏价值，但茎叶乳汁有毒，园林应用不广。

## ECOLOGICAL DISTRIBUTION

Distributed in China, India, Pakistan, Sri Lanka, Myanmar, Vietnam and Malaysia, etc.

## GROWTH HABIT

Heliophytes, suitable for growing at a temperature of 20-35℃; grow on sunny slopes, open fields and beaches.

## APPLICATION,VALUE

The whole plant can be used as green manure. After being dried, the milk can be used as a raw material for gum, and can also be used for making tannin and yellow dye. Root, stem, leaf and fruit can be used for medicinal use, with anti-inflammatory, antibacterial, phlegm and detoxification effects. The milk of stems and leaves is poisonous and contains a variety of cardiac glycosides for medicinal use; the bark can cure epilepsy. The stem bark fiber can be used to make ropes, paper, artificial cotton, woven linen, sacks, etc., and its seed hair can be used as filler and velvet raw material; the fibers can be used to replace cotton fibers in textile production. The fruit shape is peculiar, has certain ornamental value, but the milk of stems and leaves is poisonous, the garden application is not widespread.

# 04 / 铁海棠

*Euphorbia milii*
大戟科 Euphorbiaceae    大戟属 *Euphorbia*
别名：虎刺梅、虎刺、麒麟刺

## 植物形态

蔓生灌木。茎多分枝，具纵棱，密生硬而尖的锥状刺，旋转排列于棱脊上。叶互生，倒卵形或长圆状匙形，具小尖头，全缘，长1.5~5cm，宽5~20mm；托叶钻形，极细，早落。二歧状复花序，生于枝上部叶腋；柄基部具1枚膜质苞片，红色。蒴果三棱状卵形，长约3mm，宽4mm，花果期全年。种子卵柱状，直径2mm，灰褐色。

## 生态分布

原产马达加斯加，广泛栽培于热带和温带。

## PLANT MORPHOLOGY

Trailing shrub. Stems multi-branched, with longitudinal edges, dense and hard and pointed cone-like spines, arranged in rotation on the ridges. Leaves alternate, obovate or oblong spoon-shaped, with small pointed head, entire, 1.5-5cm in length and 5-20mm in width; stipules subulate, very fine, early fall. Inflorescence dichotomous complex, born in the upper leaf axils of the branches; the base of the stalk has a membranous bract, red. Capsule three-sided and ovoid, about 3mm in length and 4mm in width, flowering and fruiting period all year round. Seeds ovoid cylindrical, 2mm in diameter, grayish brown.

## ECOLOGICAL DISTRIBUTION

Native to Madagascar, widely cultivated in

## 生长习性

喜温、喜湿、喜光，不耐寒。适宜生长在疏松、排水良好的腐殖土中。

## 用途、价值

用于园林绿化。花期长，红色苞片鲜艳夺目，可作盆栽，用于宾馆、商场等公共场所装饰。根、茎、叶、乳汁均可入药，外敷可治瘀痛、骨折等。

tropical and temperate zones.

### GROWTH HABIT

Thermophilic, wet-loving, light-favored, unhardy, suitable for loose and well-drained soil.

### APPLICATION,VALUE

Used for landscaping. Long flowering period, red bracts, bright and eye-catching, used as potted plants for decoration in public places such as hotels and shopping malls. Roots, stems, leaves, and milk applicable in medicine, and external application can treat blood stasis, pain, fractures, etc.

# 05 / 黄槐决明

*Senna surattensis*
豆科 Leguminosae　决明属 *Senna*
别名：黄槐、决明子

## 植物形态

灌木或小乔木，高5~7m。分枝多；树皮光滑，灰褐色。偶数羽状复叶，长10~15cm，小叶7~9对，长椭圆形或卵形，长2~5cm，宽1~1.5cm，全缘，下面粉白色，疏被长柔毛。总状花序；花瓣鲜黄至深黄色，卵形至倒卵形。荚果扁平，带状，长7~10cm，宽8~12mm，顶端具细长的喙，花果期几乎全年。种子8~12枚。

## 生态分布

分布于中国、印度、巴基斯坦、斯里兰卡、印度尼西亚、菲律宾和澳大利

## PLANT MORPHOLOGY

Shrub or small tree. Plant height 5-7m. Many branches; smooth bark, gray-brown. Even-pinnate compound leaves, 10-15cm in length, 7-9 pairs of lobules, oblong or ovate, 2-5cm in length, 1-1.5cm in width, entire, powdery white underneath, sparsely pilose. Inflorescence racemose petals bright yellow to deep yellow, ovate to obovate. Pod flat, band-shaped, 7-10cm in length and 8-12mm in width, with a slender beak at the top, flowering and fruiting period almost all year round. Seeds 8-12 pieces.

## ECOLOGICAL DISTRIBUTION

Distributed in China, India, Pakistan, Sri Lanka, Indonesia, Philippines, Australia, etc.

亚等地。

## 生长习性

　　幼树耐阴，成年树喜光。能耐短期-2℃低温及一般霜冻，耐干旱，不抗风，不耐积水，对土壤水肥条件要求不严。

## 用途、价值

　　树形优美，花期长，开花时满树黄花，宜植于庭园和绿地或植作行道树，常作绿篱和园林观赏植物。叶可药用，具有清热解毒、润肺的功效。

## GROWTH HABIT

Young trees shade-tolerant, adult trees light-favored, resistant to short-term -2°C temperature and general frost, drought resistance, wind resistance and ponding resistance, not strict to the soil and water quality.

## APPLICATION,VALUE

Beautiful tree shape, long flowering period, full of yellow flowers when it blooms. Suitable for planting in garden and green space or as street tree, often as hedge and garden ornamental plant. leaves applicable in medicine, cool, detoxify and moisten the lung.

# 06 / 骆驼刺

*Alhagi sparsifolia*
豆科 Leguminosae　骆驼刺属 *Alhagi*
别名：骆驼草

## 植物形态

半灌木。高20~60cm；茎直立，具细条纹，幼茎具短柔毛，基部分枝。叶互生，卵形、倒卵形或卵圆形，长8~15mm，宽5~10mm。总状花序，花序轴变成坚硬的锐刺，刺长为叶的2~3倍，刺上具3~6花；花长5~10mm；苞片钻状，长约1mm；花萼钟状，长4~10mm，被短柔毛；花冠深紫红色，旗瓣先端钝圆或截平，翼瓣长圆形，龙骨瓣与旗瓣约等长。荚果线形，常弯曲。

## 生态分布

原产中国。分布于哈萨克斯坦、乌

## PLANT MORPHOLOGY

Subshrub. Plant height 20-60cm; stem erect, with fine stripes, young stem pubescent, base branched. Leaves alternate, ovate, obovate or ovoid, 8-15mm in length and 5-10mm in width. Racemose, inflorescence axis into hard sharp spines, spines 2-3 times as long as leaves, with 3-6 flowers on the spines; flowers 5-10mm in length; bracts subulate, about 1mm in length; calyx campanulate, 4-10mm in length, pubescent; corolla dark purple, flag petal apex obtuse or truncate, wing petal oblong, keel petal about the same length as the flag petal. Pods linear, often curved.

## ECOLOGICAL DISTRIBUTION

Native to China. Distributed in Kazakhstan, Uzbekistan, Pakistan, Kyrgyzstan and Tajikistan, etc.

兹别克斯坦、巴基斯坦、吉尔吉斯斯坦和塔吉克斯坦等地。

## 生长习性

耐旱，耐盐碱，抗涝，适应力强。常见于戈壁滩、沙漠。

## 用途、价值

是一种防风固沙植物。根系发达，可吸收盐分，改良土壤，在绿洲外围建设以骆驼刺为优势种的防护林及盐化草甸，对于抑制草场退化、减轻干旱荒漠农区绿洲的盐渍化及沙化、保护及扩大绿洲等起着重要作用。根系是骆驼在沙漠中的食物补充物，是一种良好的饲草和青贮饲料。刺叶分泌的液汁可药用。

## GROWTH HABIT

Drought resistance, salt-alkali resistance and flood-enduring, strong adaptability to survival. common in gobi desert and desert.

## APPLICATION,VALUE

Wind-proof and sand-fixing plant. Playing an important role in restraining grassland degradation, alleviating the salinization and desertification of oasis in arid desert agricultural area, protecting and expanding oasis and so on. Root system is the food supplement of camel in desert, good forage grass and silage. The juice secreted by thorn leaves applicable in medicine.

# 07 / 金合欢

*Acacia farnesiana*

豆科 Leguminosae　相思树属 *Acacia*

别名：消息花、刺毡花、鸭皂树

## 植物形态

　　灌木或小乔木，高1~4m。树皮粗糙，褐色；多分枝，小枝常呈"之"字形弯曲，有小皮孔。托叶针刺状，刺长1~2cm，生于小枝上的较短；二回羽状复叶，叶轴槽状，被灰白色柔毛，有腺体；小叶线状长圆形，长0.2~2cm，宽0.1~0.5cm，无毛。头状花序簇生于叶腋；总花梗被毛，长1~3cm；苞片位于总花梗的顶端或近顶部；花黄色，有香味；花萼长0.2cm，5齿裂；花瓣连合呈管状；雄蕊长约为花冠的2倍；子房圆柱状，被微柔毛；花期3~6月。种子多枚，褐色，卵形；荚果膨胀，近圆柱

## PLANT MORPHOLOGY

　　Shrub or small tree. Plant height 2-4m. Bark rough, brown; multi-branched, the twigs often bent in a zigzag shape with small bark holes. Stipules needle-shaped, with spines 1-2cm long, and shorter on twigs; two pinnately compound leaves, with axillary leaf shafts, grayish pilose, glandular; leaflets linear-oblong, 0.2-2cm long, 0.1-0.5cm wide, glabrous. Heads clustered in axils of leaves; pedicels hairy, 1-3cm long; bracts located at or near the top of the total pedicel; flowers yellow and fragrant; calyx 0.2cm, 5-toothed; petals congeal tubular; stamens about 2 times longer than corolla; ovary terete, puberulent; flowering period from March to June. Seeds numerous, brown, ovate; pod inflated, nearly cylindrical, brown, glabrous, stiff straight or curved; fruiting period from July to November.

状，褐色，无毛，劲直或弯曲；果期7~11月。

## 生态分布

原产澳大利亚。分布于中国、印度、巴基斯坦等地。

## 生长习性

喜温暖湿润，耐干旱，喜光。以肥沃、湿润的微酸性土壤为最适合，要求土壤疏松肥沃、腐殖质含量高。

## 用途、价值

木材坚硬，可用于制作家具、室内装饰材、车船、农具等；造纸材可生产书写纸、印刷纸；枝桠为可再生能源薪材、炭材，也可培菇。全株可作药用，根入药能清热，具祛痰、消炎等作用；根及荚果含单宁，可为黑色染料，果可作为饲料；叶可作饲料、绿肥、野菜；种子可食用、饲用、油用、胶用。花香，供提取香精，可提炼芳香油作高级香水等化妆品的原料；茎流出的树脂可供美工用；树皮可用于提取栲胶、单宁胶、纤维；树胶广泛用于胶水、乳化剂、墨水、印染、糖果、制药等工业。

可制作直干式、斜干式、双干式、丛林式、露根式等多种不同的盆景；也可作绿篱。环保树种，观赏树种。

## ECOLOGICAL DISTRIBUTION

Native to Australia. Distributed in China, India, Pakistan, etc.

## GROWTH HABIT

It likes warm and humid environment and resistant to drought. It delights the light and the sun. The soil is most suitable for fertile and moist slightly acidic soil, which requires loose and fertile soil and high humus content.

## APPLICATION,VALUE

The wood is hard and can be used for furniture, interior decoration materials, vehicles and boats, agricultural tools, etc; paper-making materials can produce writing paper and printing paper; the twigs are renewable energy fuel wood, charcoal wood and mushrooms. The whole plant can be used for medicinal purposes, the root can be used to clear the heat, and can eliminate phlegm and inflammation. the roots and pods contain tannin, which can be black dye, and the fruit can be used as feed; the leaves can be used as feed, green manure, and wild vegetables; the seeds are edible, feed, oil and rubber. The flower is extremely fragrant, for extracting essences, and can be used to extract aromatic oils as raw materials for high-grade perfumes and other cosmetics; the resin flowing out of the stem can be used for arts and crafts; bark can be used to extract tannin, tannin, and fiber; gum is widely used in industries such as glue, emulsifier, ink, printing and dyeing, candy, and pharmaceuticals.

It can produce a variety of different bonsais such as direct-drying, oblique-drying, double-drying, jungle-type, dew-rooting, etc; it can also be used as hedges, environmental protection tree species, and ornamental tree species.

# 08 / 长春花

## 植物形态

亚灌木。高15~60cm，枝条有水液，灰绿色，具条纹。叶膜质，倒卵状长圆形，长3~4cm，宽1.5~2cm，有短尖头。花2~3朵，呈聚伞花序；花萼5深裂，花冠红色，高脚碟状，花冠筒圆筒状，内面具疏柔毛，喉部紧缩，具刚毛。蓇葖双生，长约25mm，直径3mm，外果皮厚纸质，有条纹，被柔毛；花果期全年。种子黑色，圆筒形，两端截形。

## 生态分布

原产非洲东部。现栽培于各热带和亚热带地区。

## 生长习性

喜高温、高湿、耐半阴，不耐寒，不耐盐碱。最适宜温度为20~35℃，适生于排水良好、通风透气的砂质或富含腐殖质的土壤。

## 用途、价值

可作为观赏植物，常种植于花坛边缘。含长春碱，可药用，有降血压之效。

## PLANT MORPHOLOGY

Subshrub. Plant height 15-60cm, the branches contain water, grayish green and striped. Leaves membranous, obovate oblong, 3-4cm in length, 1.5-2cm in width, with short pointed head. 2-3 flowers in cymes; calyx 5-lobed, corolla red, high foot saucer shaped, corolla tube cylindric, inner surface sparsely pilose, throat constricted, setose. Follicles twins, about 25mm in length and 3mm in diameter. Exocarp thick, papery, striate and pilose. Flowering and fruiting period all year round. Seeds black, cylindric, truncate at both ends.

## ECOLOGICAL DISTRIBUTION

Native to eastern. Cultivated in tropical and subtropical regions.

## GROWTH HABIT

Favoring heat and high humid, half shade resistance, but not cold, salt and alkali resistance. Suitable for temperature is 20-35℃. Favored growing in sandy soil with good drainage and ventilation or rich in humus.

## APPLICATION,VALUE

As an ornamental plant, it is often planted on the edge of flower bed. Contains vinblastine, which can be used as medicine and effective in lowering blood pressure.

# 09 / 扶桑

*Hibiscus rosa-sinensis*
锦葵科 Malvaceae　木槿属 *Hibiscus*
别名：状元红、桑槿、大红花

## 植物形态

  常绿灌木，高1~3m；小枝圆柱形，疏被星状柔毛。叶阔卵形或狭卵形，长4~10cm，宽2~5cm，先端渐尖，基部圆形或楔形，边缘具粗齿或缺刻，两面除背面沿脉上有少许疏毛外均无毛；叶柄长0.5~2.0cm，上面被长柔毛；托叶线形，长0.5~1.0cm，被毛。花单生于上部叶腋间，常下垂；花梗长3~8cm，疏被星状柔毛或近平滑无毛，近端有节；萼钟形，长约2cm，被星状柔毛，卵形至披针形；花冠漏斗形，玫瑰红或淡红、淡黄等色；花瓣倒卵形，先端圆，外面疏被柔毛；雄蕊柱长3~9cm，平滑无

## PLANT MORPHOLOGY

  Evergreen shrub, about 1-3m high; branchlets cylindrical, sparsely stellate pilose. Leaves broadly ovate or narrowly ovate, 4-10cm long, 2-5cm wide, tapering at the apex, round or wedge-shaped at the base, with thick teeth or notches on the edges, except for a few sparse hairs along the veins on both sides hairless; petiole 5-20mm long, villous above; stipules linear, 5-10mm long, hairy. Flower solitary in the axilla of the upper leaf, often pendulous; pedicel 3-8cm long, sparsely stellate pilose or sub-smooth glabrous, with proximal nodes; calyx bell-shaped, about 2cm long, stellate pilose, ovate to lanceolate; corolla funnel-shaped, rose red or light red, light yellow and other colors; petals obovate, apex round, sparsely pubescent outside; stamens 3-9cm long, smooth and glabrous; style branches

毛；花柱5；花期全年。蒴果卵形，平滑无毛。

## 生态分布

原产中国。在巴基斯坦、马来西亚、苏丹、美国夏威夷等热带地区广泛栽培。

## 生长习性

喜温暖湿润气候和阳光充足的环境，但不耐霜冻。对土壤要求不严，在肥沃、湿润、排水良好的土壤中生长良好。

## 用途、价值

根、叶、花均可入药，有清热利水、解毒消肿之功效。花大色艳，花期长，盆栽是布置节日公园、花坛及家庭养花的最好花木之一，具有很高的观赏价值。

5; flowering throughout the year. Capsules ovate, smooth and glabrous, with a beak.

### ECOLOGICAL DISTRIBUTION

Originated in China. Widely cultivated in tropical areas such as Pakistan, Malaysia, Sudan and Hawaii.

### GROWTH HABIT

It likes a warm and humid climate, and likes a growing environment with plenty of sunlight, but not resistant to frost. The requirements for the soil are not strict, and it grows well in fertile, humid, well-drained soil.

### APPLICATION,VALUE

Its roots, leaves and flowers can be used as medicine, which has the effect of clearing away heat and promoting water, detoxifying and detumescence. The flowers are bright and colorful, and the flowering period is long. Potted plants are one of the best flowers and trees for arranging festival parks, flower beds and family flowers.

# 10 / 龙船花

*Ixora chinensis*
茜草科 Rubiaceae　龙船花属 *Ixora*
别名：山丹、卖子木、蒋英木

## 植物形态

灌木。高0.8~2m。小枝初时深褐色，有光泽，老时呈灰色，具线条。叶对生，披针形或长圆状倒卵形，长6~13cm，宽3~4cm；托叶鞘状。花序顶生，多花，具短总花梗；总花梗与分枝均呈红色，基部常有2枚小型叶承托；花冠红色或红黄色，盛开时长20~30mm，顶部4裂，裂片扩展或外翻；花柱短伸出冠管外；花期5~7月。果近球形，双生，成熟时黑红色。种子长、宽4~4.5mm，上凸下凹。

## PLANT MORPHOLOGY

Shrub. Plant height 0.8-2m. Branchlets initially dark brown, glossy, gray when old, with lines. Leaves opposite, lanceolate or oblong obovate, 6-13cm in length and 3-4cm in width; stipules sheathed. Inflorescence terminal, many flowers, with short total pedicel; total pedicel and branches red, often with two small leaves at the base; corolla red or red yellow, blooming 20-30mm in length, top 4-lobed, lobes extended or reversed; style short, extending out of the crown tube; flowering period from May to July. Fruit sub globose, twin, black red at maturity. Seeds 4-4.5mm in length and 4.5mm in width, up convex and down concave.

## ECOLOGICAL DISTRIBUTION

Native to China, Myanmar and Malaysia.

## 生态分布

原产中国、缅甸和马来西亚，分布于巴基斯坦、越南、菲律宾等地。

## 生长习性

喜高温、喜光、喜湿，不耐低温，适生温度20~35℃。适宜生长在富含有机质的酸性砂质壤土或腐殖质壤土上。

## 用途、价值

在园林中应用广泛，可作切花；适宜露地栽植，应用在庭院、道路旁及各风景区。根、茎药用，可清热凉血、活血止痛；花主治月经不调、高血压等症。

Distributed in Pakistan, Vietnam, Philippines, etc.

### GROWTH HABIT

Favoring heat, sunny and humid, and not low temperature resistance. Suitable for temperature is 20-35°C. Favored growing in acid sandy loam or humus loam rich in organic matter.

### APPLICATION,VALUE

It is widely used in gardens and can be used as cut flowers. It is suitable for open planting and used in courtyard, roadside and scenic spots. Root and stem are applicable in medicine for clearing heat and cooling blood, promoting blood circulation and relieving pain; flowers are mainly used for irregular menstruation and hypertension.

# 11 / 夹竹桃

夹竹桃科 Apocynaceae  夹竹桃属 *Nerium*

*Nerium oleander*

别名：红花夹竹桃

## 植物形态

　　常绿灌木。高1~5m。枝条灰绿色，含水液；嫩枝条具棱，被微毛。叶披针形，轮生，枝下部叶为对生，叶缘反卷，叶背有多数洼点；叶柄具腺体。聚伞花序，顶生，着花数朵；花萼5深裂，红色，内面基部具腺体；花冠为漏斗状，深红色或粉红色；花期全年。蓇葖果，离生，平行或并连，长圆形，长10~23cm，直径6~10mm，绿色，具细纵条纹，栽培少结果。种子长圆形，种皮被锈色短柔毛，顶端具黄褐色绢质种毛。

## PLANT MORPHOLOGY

　　Evergreen shrub. Plant height 1-5m. Branches greyish green and watery; twigs angledand puberulent. Leaves lanceolate, whorled, and the lower branches opposite. Leaf margin reversed, with many pits on the back of the leaf; petiole having glands. Cymes, terminal, with several flowers; calyx 5-lobed, red, inner base with glands; corolla funnel-shaped, dark red or pink; flowering period all year-round. Follicles, free, parallel or connected, oblong, 10-23cm in length, 6-10mm in diameter, green, with thin longitudinal stripes, less fruit in cultivation. Seeds oblong, testa covered with rust colored pubescence, top with yellowish brown silky seed hairs.

## 生态分布

原产伊朗、印度等地。现广植于亚热带及热带地区。

## 生长习性

喜温，喜光，耐阴，不耐寒。适生于排水良好、肥沃的中性土壤。

## 用途、价值

用于园林绿化。叶片有抗烟雾、抗灰尘、抗毒物和净化空气、保护环境的能力，被人们称为"环保卫士"。药用，用于治疗心力衰竭、癫痫等症；植株毒性强，勿食用，可作杀鼠药及堕胎药。茎皮纤维为优良混纺原料。

## ECOLOGICAL DISTRIBUTION

Native to Iran, India, etc. Widely planted in subtropical and tropical areas.

## GROWTH HABIT

Light-favored and thermophilus plant, shade tolerance, but not cold-resistant. Favored growing in neutral soil with good drainage and fertility.

## APPLICATION,VALUE

For landscaping. It is called "environmental protection guard", because the leaf has the ability of resisting smoke, dust, poison, purifying air and protecting environment. Applicable in medicine, effective in heart failure, epilepsy and other diseases; plants are highly toxic and should not be eaten, but can be used as rodenticide and abortion drugs. Stem skin fiber is an excellent blended material.

# 12 / 黄花夹竹桃

夹竹桃科 Apocynaceae

*Thevetia peruviana*
黄花夹竹桃属 *Thevetia*
别名：黄花状元竹

## 植物形态

常绿灌木。高1~5m。树皮棕褐色，皮孔明显，全株具乳汁。叶互生，革质，线状披针形或线形，长8~15cm，宽1~2cm。顶生聚伞花序，黄色，具香味；花萼绿色，5裂，裂片三角形；花冠漏斗状，花冠裂片向左覆盖，比花冠筒长；花期5~12月。核果，扁三角状球形，直径2.5~4mm，内果皮木质，生时绿色，干时黑色；果期8月到翌年春季。种子2~4枚，淡灰色，长约20mm。

## 生态分布

原产美洲热带、西印度群岛及墨西

## PLANT MORPHOLOGY

Evergreen shrub. Plant height 1-5m. bark brown, with obvious lenticels, the whole plant having milk. Leaves alternate, leathery, linear lanceolate or linear, 8-15cm in length and 1-2cm in width. Terminal cyme, yellow, fragrant; calyx green, 5-lobed, lobes triangular; corolla funnel-shaped, corolla lobes covered to the left, longer than corolla tube; flowering period from May to December. Drupe, oblate triangular globose, 2.5-4mm in diameter, endocarp woody, green when born and black when dry; fruiting period from August to next spring. Seeds 2-4, light gray, about 20mm in length.

## ECOLOGICAL DISTRIBUTION

Native to tropical America, West Indies and Mexico. Cultivated in tropical and subtropical regions of the world.

哥一带。现世界热带和亚热带地区均有栽培。

## 生长习性

耐旱力强，稍耐轻霜。生长于干热地区、路旁、池边、山坡疏林下；在土壤较湿润而肥沃的地方生长较好。

## 用途、价值

株型优美、花期长、花色鲜明，可作绿篱，是园林绿化的优良树种。对二氧化硫、氯气、烟尘等有毒有害气体具有很强的抵抗力，吸收能力强，是工矿区绿化的优良树种。果仁含有黄花夹竹桃素，有强心、利尿等作用；树液和种子有毒，误食可致命。叶片可杀蛆、蝇等害虫；种子可榨油，油粕可作肥料；种子坚硬，可作镶嵌物。

## GROWTH HABIT

Strong drought resistance and slight frost resistance. It grows in dry and hot areas, roadside, pond side, hillside under sparse forest; favored growing in humid and fertile soil.

## APPLICATION,VALUE

Beautiful tree type, long flowering period and bright flower color. Used as hedge. It is an excellent tree species in landscaping. It has strong resistance to sulfur dioxide, chlorine, smoke and other toxic and harmful gases, and has strong absorption capacity. It is an excellent tree species for greening in industrial and mining areas. The fruit kernel contains oleoresin, which has the effect of strengthening heart and diuresis; tree sap and seed are poisonous and can be fatal if eaten by mistake. The leaves can kill maggots, flies and other pests; the seeds can be pressed oil, and the oil meal can be used as fertilizer; the seeds are hard and can be used as inlays.

# 13 / 无花果

*Ficus carica*
桑科 Moraceae　榕属 *Ficus*
别名：红心果

## 植物形态

落叶灌木。高2~10m。树皮灰褐色，皮孔明显。叶互生，厚纸质，广卵圆形，长宽近相等，10~20cm，3~5裂，边缘具不规则钝齿，表面粗糙，背面密生细小钟乳体及灰色短柔毛；叶柄长2~5cm，粗壮；托叶卵状披针形，红色。雌雄异株；隐头花序，梨形，单生于叶腋，直径3~5cm，顶部下陷，成熟时紫红色或黄色；瘦果透镜状；花果期5~7月。

## 生态分布

分布于中国、巴基斯坦、西班牙、

## PLANT MORPHOLOGY

Deciduous shrub. Plant height 2-10m. bark grayish brown with obvious lenticels. Leaves alternate, thick papery, broadly ovate, nearly equal in length and width, 10-20cm, 3-5 lobed, margin with irregular obtuse teeth, rough surface, densely covered with small stalactites and gray pubescence on the back; petiole 2-5cm in length, stout; stipules ovate lanceolate, red. Dioecious; hypanthium, pyriform, solitary in axils of leaves, 3-5cm in diameter, sunken at the top, purplish red or yellow at maturity; achene lenticular; flowering and fruiting period from May to July.

## ECOLOGICAL DISTRIBUTION

Distributed in China, Pakistan, Spain, France, Egypt, Turkey, Afghanistan, etc.

法国、埃及、土耳其、阿富汗等地。

## 生长习性

喜温、喜湿、喜光，耐贫瘠，抗旱，不耐寒，不耐涝；适生温度5~20℃。适生于土层深厚、疏松肥沃、排水良好的砂质壤土。

## 用途、价值

榕果味甜可食或作蜜饯，又可作药用；产量高，是目前世界上投产最快的果树之一。是庭院、公园的观赏树木；叶片大，具有良好的吸尘效果，与其他植物配置栽植，可形成良好的防噪声屏障；能抵抗有毒气体和大气污染，是化工污染区绿化树种。适应性强，抗风、耐旱、耐盐碱，在干旱的沙荒地区栽植，具有防风固沙、绿化荒滩的作用。

## GROWTH HABIT

Thermophilic and wet-loving, and light-favored, barren, drought resistant, cold resistant and waterlogging resistant. Suitable for 5-20°C temperature. It is suitable for sandy loam with deep, loose and fertile soil and good drainage.

## APPLICATION,VALUE

Ficus carica is sweet and edible, or can be used as candied fruit, medicine; with high yield, it is one of the fastest fruit trees in the world. Ornamental trees in courtyards and parks; large leaves have good dust absorption effect, and can form a good noise barrier when planted with other plants. It can resist toxic gas and air pollution, and is a greening tree species in chemical pollution areas. Strong adaptability, wind resistance, drought resistance, salt and alkali resistance, planting in arid desert areas, with the role of wind and sand fixation, greening desert.

# 14 / 洋金凤

*Caesalpinia pulcherrima*
豆科 Leguminosae　云实属 *Caesalpinia*
别名：金凤花

## 植物形态

大灌木或小乔木。枝绿色或粉绿色，散生疏刺。二回羽状复叶，长10~25cm；羽片4~8对，长5~12cm；小叶7~11对，长1~2cm，先端凹缺。总状花序，近伞房状；花托凹陷成陀螺形；花瓣橙红或黄色，边缘皱波状；花丝红色，远伸出花瓣外；花柱长，橙黄色。荚果窄而薄，倒披针状长圆形，长6~10cm，宽1.5~2cm；端有长喙，成熟时黑褐色，有6~10枚种子；花果期几乎全年。

## 生态分布

原产西印度群岛。分布于中国、巴基斯坦、牙买加、古巴等地。

## 生长习性

喜高温高湿，不耐寒，喜光，不耐荫蔽。对土壤的要求不严，适生于酸性砂质土或黏土。

## 用途、价值

花冠橙红色，边缘金黄色，是观赏树木。茎、种子入药，有活血通经的功效。

## PLANT MORPHOLOGY

Large shrub or small tree, branches green or pink green, sparsely spiny. Secondary pinnate compound leaves, 10-25cm in length; 4-8 pairs of pinnae, 5-12cm in length; 7-11 pairs of leaflets, 1-2cm in length, with concave apex. Inflorescence racemose, nearly corymbose; receptacle depressed into top shape; petals orange red or yellow, edge wrinkled; filaments red, extending out of petals; style long, orange yellow. The pod is narrow and thin, oblong, 6-10cm long and 1.5-2cm wide; it has a long beak at the end and is dark brown at maturity with 6-10 seeds; the flowering and fruiting period is almost all year round.

## ECOLOGICAL DISTRIBUTION

Native to the West Indies. Distributed in China, Pakistan, Jamaica, Cuba, etc.

## GROWTH HABIT

Favoring high temperature and humidity, likes sunshine, is not cot-resistant and shade. It is suitable for acid sandy soil or clay.

## APPLICATION, VALUE

The corolla is orange red and the edge is golden yellow. It is an ornamental tree. The stem and seed can be used as medicine to activate blood circulation and dredge meridians.

# 15 / 苦郎树

唇形科 Lamiaceae　大青属 *Clerodendrum*

*Clerodendrum inerme*

别名：苦蓝盘、假茉莉

## 植物形态

直立或攀缘灌木，高0.5~2m。幼枝四棱，黄灰色，被短柔毛。叶对生，薄革质、卵形、椭圆形或椭圆状披针形、卵状披针形，长3~7cm，两面疏被黄色腺点，微反卷。聚伞花序，稀二歧分枝；花萼钟状，被柔毛；花冠白色，5裂，冠筒疏被腺点；花丝紫红色，细长，与花柱同伸出花冠。核果倒卵形，直径7~10mm，略有纵沟，多汁液，外果皮黄灰色，花萼宿存；花果期3~12月。

## PLANT MORPHOLOGY

Erect or climbing shrub. Plant height 0.5-2m. Young branches quadrangular, yellowish gray, pubescent; leaves opposite, thinly leathery, ovate, elliptic or elliptic lanceolate, ovate lanceolate, 3-7cm in length, sparsely yellow glandular dots on both sides, slightly revolute. Cymes, sparse dichotomous branches; calyx campanulate, pilose; corolla white, 5-lobed, crown tube sparsely glandular dots; filaments purplish red, slender, extending corolla with style. Drupe obovate, 7-10mm in diameter, slightly longitudinal groove, juicy, yellow gray exocarp, persistent calyx; flowering and fruiting period from March to December.

## 生态分布

分布于中国、印度、巴基斯坦等地。

## 生长习性

喜高温、喜光。适生温度20~30℃。适生于排水良好的疏松土壤。

## 用途、价值

可作为沿海防沙造林树种。根可入药，有清热解毒、舒筋活络的功效；枝叶有毒，可杀虫。

## ECOLOGICAL DISTRIBUTION

Distributed in China, India, Pakistan, etc.

## GROWTH HABIT

Light-favored. Suitable for 20-30°C temperature. It is suitable for loose soil with good drainage.

## APPLICATION, VALUE

It can be used as a coastal afforestation tree species for sand prevention. Roots can be used as medicine, and have the effects of clearing away heat and toxic materials, relaxing tendons and activating collaterals; branches and leaves are poisonous and can kill insects.

# 16 / 棉花

*Gossypium hirsutum*
锦葵科 Malvaceae  棉属 *Gossypium*

## 植物形态

　　植株灌木状，在热带地区栽培可高达6m。叶阔卵形，直径4~15cm，基部心形，常3浅裂，少为5裂，沿叶脉被粗毛，下面疏被长柔毛；叶柄长3~15cm。花单生于叶腋；小苞片具腺体，边缘具7~9齿，被长硬毛和纤毛；花白色或淡黄色，后变淡红色或紫色，长2~3cm；花期夏秋季。蒴果卵圆形，长3~5cm，具喙。种子分离，具白色长棉毛和灰白色不易剥离的短棉毛。

## 生态分布

　　原产墨西哥。分布于中国、印度、

## PLANT MORPHOLOGY

　　Plant shrubby and can be as high as 6m in tropical areas. Leaves broad ovate, 4-15cm in diameter, heart-shaped at the base, often 3-lobed, rarely 5-lobed, coarsely hairy along the vein of the leaf, sparsely villous below; petiole 3-15cm long. Flowers solitary in axils; bracteoles glandular, margin 7-9-toothed, hirsute and ciliate; flowers white or yellowish, later reddish or purple, 2-3cm in length; flowering period in summer and autumn. Capsule ovate, 3-5cm in length. Seeds separated, with white long hairs and grayish-white short cotton hairs not easily peeled.

## ECOLOGICAL DISTRIBUTION

　　Native to America and Mexico. Distributed in China, India, Pakistan, Egypt, etc.

巴基斯坦、埃及等地。

## 生长习性

喜光作物，适宜在较充足的光照条件下生长。

## 用途、价值

产量高，生产成本低，可制织物，是最重要的纤维作物、油料作物、粮食作物，也是精细化工原料和重要的战略物资，是世界上最主要的农作物之一。

## GROWTH HABIT

Light-favored crop, which is suitable for growing under sufficient light conditions.

## APPLICATION,VALUE

It is the most important fiber crop, oil crop, grain crop, fine chemical raw materials and important strategic materials, and it can be used to make fabrics, which is one of the most important crops in the world.

# 17 / 红雀珊瑚

## 植物形态

直立亚灌木。高30~80cm。茎枝粗壮，带肉质，"之"字形扭曲。叶肉质，卵形或长卵形，长2~8cm，宽2~7cm，幼叶两面被短柔毛；中脉在背面强烈凸起；托叶为一圆形的腺体。聚伞花序，每一聚伞花序为一总苞所包围；总苞鲜红或紫红色；每花仅具1雄蕊；雌花着生于总苞中央而斜伸出于总苞之外；花期12月至翌年6月。

## 生态分布

原产美洲西印度群岛。分布于中国、巴基斯坦等地。

## 生长习性

喜温，耐半阴，温度应保持在8℃以上。适生于阳光充足且通风良好的地区。适宜疏松肥沃、排水良好的土壤。

## 用途、价值

树形似珊瑚，故称"红雀珊瑚"，茎枝和叶片深绿，姿态优美，终年常青，可用于绿化栽培，适于盆栽装饰书桌、几案等。全草入药，可清热解毒，消肿止血。

## PLANT MORPHOLOGY

Erect subshrub. Plant height 30-80cm. The stems and branches are thick and fleshy, and zigzag twisted. The leaves are succulent, lamellae ovate or long ovate, 2-8cm long and 2-7cm wide, pubescent on both sides of young leaves; midrib strongly convex on the back; stipules a round gland. Cyme, each cyme is surrounded by an involucre; the involucre is bright red or purplish red; each flower has only one stamen; the female flower is placed in the center of the involucre and extends out of the involucre obliquely; the flowering period is from December to June of the following year.

## ECOLOGICAL DISTRIBUTION

Native to the West Indies. Distributed in China, Pakistan, etc.

## GROWTH HABIT

The temperature should be kept above 8°C. It is suitable for areas with plenty of sunshine and good ventilation. It is suitable for loose and fertile cultivation soil with good drainage.

## APPLICATION,VALUE

The tree looks like a coral, so it is called "Cardinal Coral". Its stems, branches and leaves are dark green, and its posture is beautiful. It can be used for greening cultivation and is suitable for potted plants to decorate desks and cases. The whole herb used as medicine can clear away heat and toxic materials, reduce swelling and stop bleeding.

## 18 / 杜鹃

## 植物形态

落叶灌木。高0.5~2m。枝条、叶柄密被亮棕褐色扁平糙伏毛。叶近革质，常集生枝端，卵圆形或披针形，长1~5cm，宽1~3cm，具细齿，叶两面疏被糙伏毛。花簇生枝顶；花萼5深裂，花冠阔漏斗形，玫瑰色、鲜红色或暗红色，上部裂片具深红色斑点；雄蕊长约与花冠相等，花柱伸出花冠外；花期4~5月。蒴果卵球形，直径0.5~1cm，密被糙伏毛，花萼宿存；果期6~8月。

## 生态分布

原产东亚。分布于中国、日本、老

## PLANT MORPHOLOGY

Deciduous shrub. Plant height 0.5-2m. Branches and petioles densely covered with bright brown flat strigose. Leaves are nearly leathery, often clustered at the end of branches, oval or lanceolate, 1-5cm in length and 1-3cm in width, with fine teeth, and leaves sparsely strigose on both sides. Flowers clustered on top of branches; calyx 5-lobed, corolla wide funnel-shaped, rose, bright red or dark red, upper lobes with dark red spots; stamens about the same length as corolla, and styles extend beyond the corolla; flowering period from April to May. Capsule ovoid, 0.5-1cm in diameter, densely strigose, persistent calyx; fruiting period from June to August.

## ECOLOGICAL DISTRIBUTION

Native to East Asia. Distributed in China,

挝、缅甸、泰国、巴基斯坦等地。

## 生长习性

喜凉爽、湿润、通风的半阴环
境，忌烈日暴晒，不耐寒，适生温度
15~25℃。适生于排水良好的酸性土壤。

## 用途、价值

枝繁叶茂，耐修剪，根桩奇特，是
优良的盆景材料；还可经修剪培育成各
种形态，是花篱的良好材料。全株可药
用，有行气活血、补虚等功效。叶、花
可入药或提取芳香油；部分花可食用。

Japan, Laos, Myanmar, Thailand, Pakistan, etc.

### GROWTH HABIT

It likes cool, humid and ventilated semi shade
environment, avoids hot sun exposure, and is not
resistant to cold. The suitable temperature for growth
is 15-25℃. It is suitable for acid soil with good
drainage.

### APPLICATION,VALUE

It is an excellent bonsai material with luxuriant
foliage and peculiar root stump, which can also
be pruned into various forms, which is a good
material for flower hedgerow. The whole plant
can be used for medicine, and has the effects of
promoting Qi, promoting blood circulation and
tonifying deficiency. Leaf and flower can be used as
medicine or extract aromatic oil; some flowers can
be edible.

# 19 / 绣球

## 植物形态

灌木，高达4m，树冠球形。小枝粗，无毛。叶倒卵形或宽椭圆形，长6~15cm，先端骤尖，具短尖头，基部钝圆或宽楔形，具粗齿，两面无毛或下面中脉两侧疏被卷曲柔毛；脉腋有髯毛，侧脉6~8对；叶柄粗，长1~3.5cm，无毛。伞房状聚伞花序近球形或头状，径8~20cm，分枝粗，近等长，密被紧贴柔毛；花密集。幼果陀螺状，连花柱长约4.5mm，顶端突出部分长约1mm；花期6~8月。

## 生态分布

原产东亚、美国东南部和南美等地，主要分布于中国、朝鲜、日本、美国、墨西哥等地，目前在全球大部分地区广泛栽培。

## 生长习性

喜温暖、湿润和半阴环境，适生温度18~28℃，土壤以疏松、肥沃和排水良好的砂质壤土为好。土壤pH的变化，会使绣球的花色变化较大。

## 用途、价值

在盆花、切花和园林绿化中广泛应用。

## PLANT MORPHOLOGY

Shrub, up to 4m tall, with a spherical crown. Branchlets thick, hairless. Leaves obovate or broadly elliptical, 6-15cm in length, apex abruptly pointed, with a short pointed head, base obtuse or broadly cuneate, with coarse teeth, glabrous on both sides or sparsely curly pubescent on both sides of the lower midrib, bearded hairs in the vein axils, 6-8 pairs of lateral veins; petiole thick, 1-3.5cm in length, glabrous. Umbrella shaped cymes nearly spherical or head shaped, with a diameter of 8-20cm. branches thick, nearly equal in length, densely covered with pubescence; flowers dense. Young fruit turbinate, about 4.5mm in length along with style, protruding tip about 1mm long; flowering period from June to August.

## ECOLOGICAL DISTRIBUTION

Native to East Asia, southeastern United States, and South America, mainly distributed in countries such as China, North Korea, Japan, the United States, Mexico, and is currently widely cultivated in most regions around the world.

## GROWTH HABIT

Suitable for growing in warm, humid, and semi shaded environments, with a suitable temperature of 18-28°C. The soil should be loose, fertile, and well drained sandy loam. However, changes in soil pH result in significant changes in the color of hydrangea flowers.

## APPLICATION,VALUE

Widely used in potted flowers, cut flowers, and landscaping.

# 20 / 一叶萩

*Flueggea suffruticosa*
叶下珠科 Phyllanthaceae　白饭树属 *Flueggea*
别名：叶底珠

## 植物形态

灌木。高1~3m。多分枝；小枝浅绿色，有棱槽。叶纸质，椭圆形或长椭圆形，长1~8cm，宽1~3cm，全缘或间有不整齐波状齿或细齿，托叶卵状披针形，宿存。花小，雌雄异株，簇生于叶腋；花期3~8月。蒴果，三棱状扁球形，直径约5mm，熟时淡红褐色，有网纹，具宿存萼片，果期6~11月。种子卵形，褐色有小疣状凸起。

## 生态分布

分布于中国、俄罗斯、日本、巴基斯坦等地。

## PLANT MORPHOLOGY

Shrub. Plant height 1-3m. Numerous branches. Branchlets light green, grooved. Leaves papery, elliptic or long elliptic, 1-8cm in length, 1-3cm in width, entire or intermittently with irregular undulate teeth or fine teeth, stipules ovate lanceolate, persistent. Flowers small, dioecious, clustered in leaf axils; flowering period from March to August. Capsule, trigonous oblate, 5mm in diameter, reddish brown when ripe, reticulate, with persistent sepals, fruiting period from June to November. Seeds ovate, brown, verrucose protuberances.

## ECOLOGICAL DISTRIBUTION

Distributed in China, Russia, Japan, Pakistan, etc.

## 生长习性

耐寒、抗旱、耐瘠薄，喜深厚肥沃的砂质壤土，但在干旱瘠薄的石灰岩山地上也可生长良好。

## 用途、价值

可配置于假山、草坪、河畔、路边等，具有良好的观赏价值。茎皮纤维坚韧，可作纺织原料，枝条可编制用具；花和叶供药用，对中枢神经系统有兴奋作用；根皮煮水，外用可治牛、马虱子等虫害。

## GROWTH HABIT

It is resistant to cold, drought and barren. It likes deep and fertile sandy loam, but it can grow well in arid and barren limestone mountains.

## APPLICATION, VALUE

In the garden, it is arranged in rockery, lawn, riverside, roadside, etc., which has good ornamental value. The fiber of the stem skin is tough, which can be used as textile material, and the branches can be used to make utensils; the flowers and leaves are used for medicine, and have an exciting effect on the central nervous system; the root bark boiled water can be used for external use to treat cattle, horse lice and other insect pests.

# 21 / 紫薇

千屈菜科 Lythraceae　紫薇属 *Lagerstroemia*

*Lagerstroemia indica*

别名：千日红、满堂红、蚊子花、紫兰花

## 植物形态

落叶灌木。高1~7m。树皮平滑，灰色或灰褐色；小枝纤细，具4棱，呈翅状。叶互生或对生，纸质，椭圆形或矩圆形。顶生圆锥花序，淡红色、白色或紫色；花瓣6片，皱缩；花期6~9月。蒴果，椭圆状球形，长1~2cm，幼时绿色或黄色，成熟或干燥时呈紫黑色；果期9~12月。种子有翅，长8mm。

## 生态分布

原产亚洲。分布于中国、巴基斯坦、菲律宾、印度等地。

## PLANT MORPHOLOGY

Deciduous shrub. Plant height 1-7m. Bark smooth, gray or grayish brown; branchlets slender, 4-ribbed, winged. Leaves alternate or opposite, papery, elliptic or oblong. Terminal panicle, light red, white or purple; petals 6, wrinkled; flowering period from June to September. Capsule, ellipsoid, 1-2cm in length, green or yellow when young, purple black when mature or dry; fruiting period from September to December. Seeds winged, 8mm in length.

## ECOLOGICAL DISTRIBUTION

Native to Asia. It is distributed in China, Pakistan, Philippines, India, etc.

## 生长习性

喜暖，喜光，略耐阴，耐干旱，耐寒。萌蘖性强。适生于深厚肥沃的砂质壤土。

## 用途、价值

用于公园、庭院绿化等，栽植于建筑物前、院落内，具有极高的观赏价值；具有较强的抗污染能力，对二氧化硫、氟化氢及氯气的抗性较强。木材坚硬、耐腐，可作农具、家具、建筑等用材。

## GROWTH HABIT

Favoring warm, like sunshine, slightly resistant to shade. It is resistant to drought and cold. It has strong sprouting ability. It is suitable for deep and fertile sandy loam.

## APPLICATION,VALUE

It can be used for park greening, courtyard greening and so on. It is planted in front of buildings and courtyards, with high ornamental value. It has strong antipollution ability and strong resistance to sulfur dioxide, hydrogen fluoride and chlorine gas. The wood is hard and corrosion-resistant, and can be used for farm tools, furniture, construction, etc.

# 22 / 木薯

## 植物形态

直立灌木。高1.5~3m。块根圆柱状。叶纸质，近圆形，长10~20cm，掌状深裂；叶柄长8~25cm，盾状着生。圆锥花序，长5~8cm，苞片条状披针形；花萼紫红色且有白粉霜；裂片长圆状披针形；花期9~11月。蒴果椭圆状，长1.5~1.8cm，直径1~1.5cm，具6条波状纵翅。种子长约1cm，种皮硬壳质，具斑纹。

## 生态分布

原产巴西。现热带地区广泛栽植。

## 生长习性

短日照热带作物，不耐阴，对光反应敏感。喜高温、不耐霜冻、喜湿润，不喜多雨，怕积水。耐贫瘠，适生于深厚肥沃、排水通气良好的土壤。

## 用途、价值

块根富含淀粉，是工业淀粉原料之一；根、茎、叶都含有毒物质，需经漂浸处理后方可食用；是常见的杂粮作物；具有多种保健功能。

## PLANT MORPHOLOGY

Erect shrub. Plant height 1.5-3m. Root tuber cylindrical. Leaves papery, suborbicular, 10-20cm in length, palmately deeply lobed; petiole 8-25cm in length, shield shaped. Panicle, 5-8cm in length, bracts oblate lanceolate; calyx purplish red with powdery frost; lobes oblong lanceolate; flowering period from September to November. Capsule elliptic, 1.5-1.8cm in length and 1-1.5cm in diameter, with 6 undulate longitudinal wings. Seeds about 1cm in length, seed coat hard crusty and mottled.

## ECOLOGICAL DISTRIBUTION

Native to Brazil. Widely planted in tropical areas.

## GROWTH HABIT

Short day tropical crops, not tolerant of shade, sensitive to light. Thermophilic, not frost-resistant, wet-loving, not favoring rainy, afraid of water, barren resistance, suitable for deep and fertile soil with good drainage and ventilation.

## APPLICATION,VALUE

Root tuber rich in starch, one of the raw materials of industrial starch; root, stem and leaf contain toxic substances, need to be bleached and soaked before edible; common cereal crop; a variety of health functions.

# 23 / 马缨丹

*Lantana camara*

马鞭草科 Verbenaceae　马缨丹属 *Lantana*

别名：七变花、如意草、臭草、五彩花

## 植物形态

灌木或蔓性灌木。高0.5~2m。茎枝有短而倒钩状刺。单叶对生，卵形或卵状长圆形，长3~9cm，宽1.5~5cm，具钝齿，上面具触纹及短柔毛，下面被硬毛。花序梗粗壮；苞片披针形；花萼管状，膜质，顶端有极短的齿；花冠黄色或橙黄色，开花后不久转为深红色，全年开花。果球形，紫黑色。

## 生态分布

原产美洲热带地区。分布于中国、韩国、巴基斯坦、日本、加拿大等地。

## PLANT MORPHOLOGY

Shrubs or creeping shrub. Plant height 0.5-2m. Stems and branches short barbed spines. Leaves simple, opposite, ovate or ovate oblong, 3-9cm in length and 1.5-5cm in width, obtuse teeth, with contact lines and pubescence on the upper surface and hirsute on the bottom. Inflorescence peduncle stout; bracts lanceolate; calyx tubular, membranous, with very short teeth at the tip; corolla yellow or orange yellow, turning dark red shortly after flowering, flowering period throughout the year. Fruit spherical, purple black.

## ECOLOGICAL DISTRIBUTION

Native to tropical America. Distributed in China, South Korea, Pakistan, Japan, Canada, etc.

## 生长习性

喜温暖湿润、光照充裕的环境，耐干旱，稍耐阴，不耐寒。对土质要求不严，适生于肥沃、疏松的砂质土壤。

## 用途、价值

园林观赏植物，可作绿篱和盆栽。具有固土截流、涵养水源、改良土壤、改善生态环境的作用，是护坎、护坡、护堤的优良灌木。根、叶、花可作药用，具有清热解毒的功效。叶有杀虫作用，可用于制造生物杀虫剂。

## GROWTH HABIT

Favoring warm and humid environment with plenty of light. Drought, shade and cold resistance. Suitable for fertile and loose sandy soil.

## APPLICATION,VALUE

Garden ornamental plants, used as hedgerow and potted. Effective in soil fixation and river closure, water conservation, soil improvement and ecological environment improvement. Excellent shrub for protecting bank, slope and dike. The root, leaf and flower applicable in medicine, effective in clearing away heat and detoxification. Leaves insecticidal, used to produce biological insecticides.

## 24 / 百合竹

龙舌兰科 Agavaceae　龙血树属 Dracaena

*Dracaena reflexa*

别名：短叶竹蕉

### 植物形态

多年生常绿灌木。高2~9m；节间短。叶线形或披针形，绿色，松散成簇。花白色、较小，花序较长；花被片6，合生；花梗有关节。浆果近球形，具1~3枚种子。

### 生态分布

原产马达加斯加。分布于中国、巴基斯坦等地。

### 生长习性

喜温、喜湿、喜光照，忌夏季烈日直射。适生温度20~30℃。对土壤和肥

### PLANT MORPHOLOGY

Perennial evergreen shrub. Plant height 2-9m; internodes short; leaves linear or lanceolate, green, loosely clustered. Flowers white, small, inflorescence longer; perianth segments 6, connate; pedicel articulated. Berry subglobose, seeds 1-3.

### ECOLOGICAL DISTRIBUTION

Native to Madagascar. Distributed in China, Pakistan, etc.

### GROWTH HABIT

Thermophilic, wet-loving, light-favored, avoid direct sunlight in summer. suitable for 20-30°C temperature, not strict to the soil and fertilizer quality.

料要求不严。

## 用途、价值

耐阴性好，可盆栽置于阳台、客厅和窗台观赏，是优良的室内观叶植物。

APPLICATION,VALUE

Good shade tolerance, potted plants can be placed on the balcony, living room and windowsill to watch, excellent indoor foliage plant.

# 25 / 福禄桐

*Polyscias guilfoylei*
五加科 Araliaceae   南洋参属 *Polyscias*
别名：圆叶南洋参

## 植物形态

常绿灌木。高1~5m；茎干灰褐色，枝干皮孔明显，或具有褐色斑块。叶互生，一回羽状复叶，小叶3~4对，对生，绿色，叶宽卵形或近圆形，基部心形，叶缘有锯齿，白色镶边。伞形花序呈圆锥状，顶生，下垂。浆果，近球形，直径4~5mm。

## 生态分布

原产印度至太平洋群岛。分布于中国、巴基斯坦等地。

## 生长习性

喜温、喜湿，不耐寒。适生温度10~25℃。喜光、耐阴，忌太阳直射。适生于排水良好，富含腐殖质的砂质土壤。

## 用途、价值

园林绿化观赏树种，株形丰满，可用于景观、绿篱、盆栽。

## PLANT MORPHOLOGY

Evergreen shrub. Plant height 1-5m; stem grayish brown with obvious lenticels or brown plates. Leaves alternate, pinnate compound leaves, 3-4 pairs of leaflets, opposite, green, leaves broadly ovate or suborbicular, base heart-shaped, leaf margin serrate, white edge. Inflorescence umbels paniculate, terminal and pendulous. Berry, subglobose, 4-5mm in diameter.

## ECOLOGICAL DISTRIBUTION

Native to India - the Pacific Islands. Distributed in China, Pakistan, etc.

## GROWTH HABIT

Thermophilic, favoring wet enviroment, unhardy. Suitable for 10-25°C temperature. Light-favored, shade-tolerant, avoids direct sunlight. Suitable for sandy soil with good drainage and rich humus.

## APPLICATION,VALUE

Ornamental tree species for landscaping, full-bodied, used for landscape, hedgerow, potted.

## 26 / 鹅掌柴

*Schefflera heptaphylla*
五加科 Araliaceae　南鹅掌柴属 *Schefflera*
别名：大叶伞、鸭脚木、鸭母树、红花鹅掌柴

### 植物形态

灌木。高0.5~7m。小枝粗壮，疏生星状茸毛，髓实心。小叶7~8，纸质，长圆状披针形，中央的较大，长24cm，宽8cm，两侧的较小，长10~11cm，宽3cm，全缘；小叶柄不等长。圆锥花序，顶生，主轴几无毛；伞形花序总状排列在分枝上，苞片卵形，有短柔毛；花淡红黄色；萼倒圆锥形，无毛，全缘；花瓣5，长三角形，无毛。果实球形，直径约3mm，黑色，无毛；花果期10月。

### 生态分布

分布于中国、印度、日本、巴基斯坦、越南等地。

### 生长习性

喜温暖、湿润、半阳环境，耐瘠薄。宜生于深厚肥沃的酸性砂质壤土。

### 用途、价值

大型盆栽植物，适宜在宾馆大厅等室内摆放。叶片可吸收尼古丁、甲醛和其他有害物质。

### PLANT MORPHOLOGY

Shrub. Plant height 0.5-7m. Branchlets stout, sparsely stellate tomentose, pith solid. Leaves with 7-8 leaflets; leaflets papery, oblong lanceolate, larger in the center, 24cm in length and 8cm in width, smaller on both sides, 10-11cm in length and 3cm in width, entire; petioles unequal in length. Umbels racemose arranged on branches, bracts ovate, pubescent; flowers reddish yellow, brownish red when dry; calyx oblong, glabrous, entire; petals 5, long triangular, glabrous. Fruit spherical, about 3mm in diameter, black and glabrous; flowering and fruiting period in October.

### ECOLOGICAL DISTRIBUTION

Distributed in China, India, Japan, Pakistan, Vietnam, etc.

### GROWTH HABIT

Favoring warm, humid and half sunny environment. Barren-resistant. Favored growing in acid sandy loam, deep and fertile soil.

### APPLICATION,VALUE

Large potted plant, suitable for indoor display in hotel hall. Leaves can absorb nicotine, formaldehyde and other harmful substances.

# 27 / 红桑

*Acalypha wilkesiana*
大戟科 Euphorbiaceae　铁苋菜属 *Acalypha*
别名：铁苋菜

## 植物形态

灌木。高1~4m。叶纸质，阔卵形，长10~18cm，宽6~12cm，绿色或浅红色，常有不规则的红色或紫色斑块，边缘具粗圆锯齿；叶柄具疏毛；托叶狭三角形，具短毛。苞片卵形，苞腋具雄花9~17朵，排成团伞花序；花期全年。蒴果，直径4mm，疏生长毛。种子球形，直径2mm。

## 生态分布

原产太平洋岛屿。现广泛栽培于热带、亚热带地区。

## PLANT MORPHOLOGY

Shrub. Plant height 1-4m. Leaves papery, broadly ovate, 10-18cm in length, 6-12cm in width, green or light red, often with irregular red or purple patches, margin coarsely serrate; petiole sparsely hairy; stipules narrowly triangular, with short hairs. Bracts ovate, axillary with 9-17 male flowers, arranged in cymes; flowering throughout the year. Capsule, 4mm in diameter, sparsely long hairy. Seeds spherical, 2mm in diameter.

## ECOLOGICAL DISTRIBUTION

Native to Pacific Islands. Widely cultivated in tropical and subtropical areas.

## 生长习性

热带树种，喜高温多湿，喜光，耐高温，抗寒力低，不耐霜冻。对土壤水肥条件的要求较高，适生于疏松、排水良好的土壤。

## 用途、价值

常作庭院、公园中的绿篱和观叶灌木，可配置在灌木丛中，是热带庭园绿化的优良树种。叶可药用，用于治疗跌打损伤等。

## GROWTH HABIT

Tropical tree species, favoring heat, humid and sunny, drought resistance, but not cold and frost resistance. Strict to the soil quality of water and fertilizer. Favored growing in loose soil with good drainage.

## APPLICATION,VALUE

Used as hedgerow and foliage shrub in garden and park. It can be arranged in the bush. A good tree species for tropical garden greening. Leaves applicable in medicine, effective in injuries and other.

# 28 / 茉莉花

*Jasminum sambac*
木樨科 Oleaceae　素馨属 *Jasminum*
别名：茉莉

## 植物形态

直立或攀缘灌木。高0.5~3m。小枝被疏柔毛。叶对生，纸质，圆形或卵状椭圆形，长4~15cm，宽2~8cm，下面脉腋常具簇毛；叶柄被柔毛，具关节。聚伞花序，顶生；花序梗被短柔毛；花冠白色；花期5~8月。果球形，直径约1cm，成熟时紫黑色；果期7~9月。

## 生态分布

原产印度。分布于中国、伊朗、埃及、巴基斯坦、西班牙、法国等地。

## PLANT MORPHOLOGY

Erect or climbing shrub. Plant height 0.5-3m. Branchlets sparsely pilose. Leaves opposite, papery, round or ovate elliptic, 4-15cm in length and 2-8cm in width, often tufted in the lower vein axils; petiole pilose, articulated. Cymes, terminal; peduncle pubescent; corolla white; fruiting period from May to August. fruit spherical, about 1cm in diameter, and purple black at maturity; fruiting period from July to September.

## ECOLOGICAL DISTRIBUTION

Native to India. Distributed in China, Iran, Egypt, Pakistan, Spain, France, etc.

## 生长习性

喜温、喜湿，在通风良好、半阴环境生长最好。忌暴晒。适生于含有大量腐殖质的微酸性砂质壤土。

## 用途、价值

是常见庭园及盆栽观赏芳香花卉，叶色翠绿，花色洁白，香味浓厚，可加工成花环等装饰品。花极香，可提取茉莉花油，是著名的花茶原料及重要的香精原料；花、叶可药用，有治目赤肿痛、止咳化痰的功效。

## GROWTH HABIT

Favoring warm and humid. It grows best in well-ventilated and half shade environment. Avoid exposure to sunlight. Favored growing in slightly acid sandy loam with a lot of humus.

## APPLICATION,VALUE

A common garden and potted ornamental fragrant flowers, green leaves, white flowers, strong fragrance, can be processed into garlands and other decorations. Flowers are very fragrant and can be extracted from lily oil. It is the famous raw material and essential flavor material of scented tea. Flowers and leaves are applicable in medicine, and effective in treating red eye swelling and pain, relieving cough and resolving phlegm.

## 29 / 朱蕉

*Cordyline fruticosa*

天门冬科 Asparagaceae　朱蕉属 *Cordyline*

别名：红铁树、红叶铁树、铁莲草、朱竹、铁树

## 植物形态

直立灌木。高1~3m。叶聚生于茎或枝的上端，长圆形或长圆状披针形，长25~50cm，宽5~10cm，绿或紫红色；叶柄有槽，基部宽，抱茎。圆锥花序，长30~60cm，侧枝基部有大苞片；花淡红、青紫或黄色；外轮花被片下部紧贴内轮形成花被筒，上部盛开时外弯或反折；花期11月至翌年3月。

## 生态分布

分布于中国、巴基斯坦、印度、泰国、越南等地。

## PLANT MORPHOLOGY

Erect shrub. Plant height 1-3m. Leaves clustered on the upper end of stem or branch, oblong or oblong lanceolate, 25-50cm in length and 5-10cm in width, green or purplish red; petiole grooved, base wide, embracing stem. Panicle, 30-60cm in length, with large bracts at the base of lateral branches; flowers light red, cyan purple or yellow; lower part of outer perianth close to the inner ring to form a perianth tube, and upper part bent or reflexed when full blooming; flowering period from November to March of the following year.

## ECOLOGICAL DISTRIBUTION

Distributed in China, Pakistan, India, Thailand, Vietnam, etc.

## 生长习性

喜高温多湿气候，喜阴，不耐旱，不耐寒。适生于富含腐殖质和排水良好的酸性土壤。

## 用途、价值

株形美观，可作园林绿化和盆栽，是布置室内场所的常用植物。

Favoring heat and humid climate, shade-resistant. Not drought and cold resistance. Favored growing in acid soil with rich humus and good drainage.

Beautiful plant shape and used for landscaping and potting. A common plant for indoor layout.

# 30 / 紫穗槐

*Amorpha fruticosa*

豆科 Leguminosae 紫穗槐属 *Amorpha*
别名：槐树、紫槐、棉槐、棉条

## 植物形态

落叶灌木。高1~4m。丛生，茎灰褐色。叶互生，奇数羽状复叶，长10~15cm；小叶11~25片，卵形或椭圆形，长1~4cm，宽0.6~2cm，下面被白色短柔毛和黑色腺点。穗状花序，长7~15cm，花序梗密被短柔毛；花萼钟状；花冠紫色，旗瓣心形，基部具短瓣柄，无翼瓣与龙骨瓣。荚果，长圆形，长0.6~1cm，宽0.2~0.3cm，微弯曲，具小突尖，成熟时棕褐色；花果期5~10月。

## 生态分布

原产美国东北部和东南部。分布于

## PLANT MORPHOLOGY

Deciduous shrub. Plant height 1-4m. Tufted, stem gray brown. Leaves alternate, odd pinnate compound leaves, 10-15cm in length; 11-25 leaflets, ovate or elliptic, 1-4cm in length and 0.6-2cm in width, covered with white pubescence and black glandular dots below. Inflorescence spikes, 7-15cm in length, peduncle densely pubescent; calyx campanulate; corolla purple, flag petal heart-shaped, base with short petiole, without wing and keel. Pods, oblong, 0.6-1cm in length and 0.2-0.3cm in width, slightly curved, with small protuberances, brownish brown at maturity; flowering and fruiting period from May to October.

## ECOLOGICAL DISTRIBUTION

Native to the northeast and southeast of the

中国、巴基斯坦等地。

## 生长习性

适生温度10~16℃，要求光线充足。对土壤要求不严。耐贫瘠、耐干旱、耐盐碱、耐寒；根部固氮，对土壤具改良作用。

## 用途、价值

枝条直立匀称，树形美观，用作园林绿化。抗风力强，生长快，枝叶繁密，是防风林的首选树种；郁闭度高，截留雨量能力强，根系萌蘖性强，不易生病虫害，具有根瘤，是保持水土的优良树种。枝条用作绿肥和饲料；叶可调制成草粉，具祛湿消肿功效。

United States. Distributed in China, Pakistan, etc.

## GROWTH HABIT

Suitable for 10-16°C temperature, and sufficient light is required. The soil is not strict. It has strong resistance to barren soil, drought, salt and alkali, cold resistance, and nitrogen fixation in roots can improve soil quality.

## APPLICATION,VALUE

The branches are upright and symmetrical, and the tree shape is beautiful, which is used for landscaping. Strong wind resistance, fast growth, dense branches and leaves, is the first choice for windbreaks. It has high canopy density, strong ability to intercept rainfall, strong root sprouting, and is not easy to produce diseases and pests. It has nodules and is an excellent tree species for soil and water conservation. Branches are used as green manure and feed. Leaves can be adjusted to make grass powder, which has the effect of eliminating dampness and reducing swelling.

# 31 / 金边黄杨

*Euonymus japonicus* 'Aurea-marginatus'
卫矛科 Celastraceae　卫矛属 *Euonymus*
别名：金边冬青卫矛、金边大叶黄杨

## 植物形态

常绿灌木。高3~5m。小枝四棱，具细微皱突。叶革质，倒卵形或椭圆形，长3~5cm，宽2~3cm，边缘具浅细钝齿；叶片有较宽的黄色边缘。聚伞花序，花白绿色，直径5~7mm；花瓣近卵圆形。蒴果近球状，直径约8mm，淡红色；种子椭圆状，长约6mm，直径约4mm，假种皮橘红色。

## 生态分布

分布于中国、日本、巴基斯坦等地。

## PLANT MORPHOLOGY

Evergreen shrub. Plant height 3-5m. Branchlets quadrangular, slightly rugose. Leaves leathery, obovate or elliptic, 3-5cm in length and 2-3cm in width, and margin with shallow and fine obtuse teeth; the leaf blade with a wide yellow edge. Cymes, white green flowers, 5-7mm in diameter; petals suborbicular. Capsule subglobose, about 8mm in diameter, pale red; seeds elliptic, about 6mm in length and 4mm in diameter, aril orange red.

## ECOLOGICAL DISTRIBUTION

Distributed in China, Japan, Pakistan, etc.

## GROWTH HABIT

Light-favored. tolerant of shade. It has strong sprouting and branching ability, and is resistant to

## 生长习性

喜光、耐阴、耐旱、耐寒冷、萌芽力和发枝力强、耐修剪、耐瘠薄，适生于肥沃、湿润的微酸性土壤。

## 用途、价值

园林绿化植物，常用于门庭和中心花坛布置，也可作盆栽观赏。具有抗污染性，可以有效对抗二氧化硫，是严重污染工矿区首选的常绿植物。

pruning. It is resistant to barren soil and suitable for growing in fertile and moist slightly acid soil.

### APPLICATION,VALUE

It is a landscaping plant, which is often used in the layout of doorways and central flower beds, and can also be used for potted viewing. It has pollution resistance and can effectively fight against sulfur dioxide, so it is the first choice for evergreen plants in heavily polluted industrial and mining areas.

# 32 / 琴叶珊瑚

*Jatropha integerrima*
大戟科 Euphorbiaceae　麻风树属 *Jatropha*
别名：变叶珊瑚花、琴叶樱、南洋樱、日日樱

## 植物形态

　　常绿灌木。高1~2m。单叶互生，倒阔披针形，常丛生于枝条顶端；叶基有2~3对锐刺，叶面浓绿色，叶背紫绿色，叶柄具茸毛，近基部叶缘常具数枚疏生尖齿。聚伞花序顶生，红色，花瓣长椭圆形，具花盘；花期春季至秋季。蒴果成熟时呈黑褐色。

## 生态分布

　　原产西印度群岛。分布于中国、巴基斯坦等地。

## PLANT MORPHOLOGY

　　Evergreen shrub. Plant height 1-2m. Leaves simple, alternate, and oblanceolate, often clustered at the top of branches. leaf base with 2-3 pairs of sharp spines, leaf surface thick green, leaf back purplish green, petiole hairy, and leaf margin near the base often with several sparse sharp teeth. Cymes, terminal, red, long elliptic petals with disk; flowering period from spring to autumn. Capsule dark brown at maturity.

## ECOLOGICAL DISTRIBUTION

　　Native to the West Indies. Distributed in China, Pakistan, etc.

## 生长习性

喜高湿环境，不耐寒；喜充足的光照，稍耐半阴。适生于疏松肥沃、富含有机质的酸性砂质土壤中。

## 用途、价值

用作庭院及室内装饰，应用于园林景观；具有净化空气、美化环境、绿化城市的作用。

## GROWTH HABIT

Favoring humidity environment, not cold-resistant; like sufficient light, slightly tolerant of semi-shade. It is suitable for growing in loose and fertile acid sandy soil with rich organic matter.

## APPLICATION,VALUE

Used as courtyard and interior decoration, applied to garden landscape; It has the functions of purifying air, beautifying the environment and greening the city.

# 33 / 决明

*Senna tora*

豆科 Leguminosae    决明属 *Senna*

别名：决明子

## 植物形态

亚灌木。高达2m。羽状复叶长4~8cm，叶柄上无腺体，叶轴上每对小叶间有1棒状腺体；小叶3对，倒卵形或倒卵状长椭圆形，长2~6cm，先端圆钝而有小尖头，基部渐窄，偏斜，上面被稀疏柔毛，下面被柔毛。花瓣黄色，下面2片稍长，长1.2~1.5cm。荚果纤细，近四棱形，两端渐尖，长达15cm，宽3~4mm，膜质。种子约25枚，菱形，光亮。

## 生态分布

分布于中国、印度、巴基斯坦、斯

## PLANT MORPHOLOGY

Subshrub-like herb, up to 2m. Pinnate compound leaves 4-8cm long, no glands on the petiole, 1 rod-shaped gland between each pair of leaflets on the leaf shaft; 3 pairs of leaflets, obovate or obovate oblong, 2-6cm long, blunt apex with small pointed head, gradually narrow and oblique at base, sparsely pilose on the top, and pilose on the bottom; petals yellow, the bottom 2 pieces slightly longer, 1.2-1.5cm long; Pods slender, nearly quadrangular, with acuminate ends, up to 15cm long, 3-4mm wide, and membranous; seeds about 25, rhomboid, bright.

## ECOLOGICAL DISTRIBUTION

Distributed in China, India, Pakistan, Sri Lanka, Indonesia, the Philippines and Australia, etc.

里兰卡、印度尼西亚、菲律宾和澳大利亚等地。

## 生长习性

喜光，喜温暖湿润气候，阳光充足有利于其生长。

## 用途、价值

种子叫决明子，有清肝明目、利水通便之功效，同时还可提取蓝色染料；苗叶和嫩果可食。

## GROWTH HABIT

Favoring light, warm and humid climate, and sufficient sunshine is conducive to its growth.

## APPLICATION,VALUE

The seeds are called cassia seeds, which have the effects of clearing the liver, improving eyesight, promoting water and laxatives, and can also extract blue dye; the leaves and tender fruits are edible.

# 34 / 使君子

*Quisqualis indica*

使君子科 Combretaceae　使君子属 *Quisqualis*

别名：四君子、史君子

## 植物形态

攀缘状灌木。高2~8m。小枝被棕黄色短柔毛。叶对生，膜质，卵形或椭圆形，长5~11cm，宽25~55mm。顶生穗状花序，组成伞房花序式；花瓣5，长1~4cm，宽3~12mm，先端钝圆，初为白色，后转淡红色；花期初夏。果卵形，长2~4cm，直径1~3cm，具明显的5条锐棱角，成熟时外果皮脆薄，呈青黑色或栗色；果期秋末。种子白色，圆柱状纺锤形。

## 生态分布

分布于中国、印度、巴基斯坦、缅

## PLANT MORPHOLOGY

Climbing shrub. Plant height 2-8m. Branchlets brownish yellow pubescent. Leaves opposite, membranous, ovate or elliptic, 5-11cm in length and 25-55mm in width. Inflorescence spikes, terminal, forming corymbs. Petals 5, 1-4cm in length, 3-12mm in width, apex blunt round, white at first, then turn light red; flowering period in early summer. Fruit ovate, 2-4cm in length and 1-3cm in diameter, with 5 sharp angles. pericarp crisp and thin at maturity, green black or chestnut color. fruiting period end of autumn. Seeds white, cylindric spindle shaped.

## ECOLOGICAL DISTRIBUTION

Distributed in China, India, Pakistan, Myanmar, Philippines, etc.

甸、菲律宾等地。

## 生长习性

喜温，根系分布广面深；宜栽于向阳背风处；对土质要求不严，适生于排水良好的肥沃砂质壤土。

## 用途、价值

树形优雅，花朵轻盈，可作观赏绿化植物。种子药用，具有驱蛔虫、蛲虫和抗皮肤真菌等功效。

## GROWTH HABIT

Favoring wet environment. The root system is widely distributed and deep. It is suitable to be planted in sunny and leeward places, and suitable for fertile sandy loam with good drainage.

## APPLICATION,VALUE

Elegant tree shape and light flowers can be used as ornamental greening plants. Seeds are used medicinally, and have the effects of expelling ascaris, pinworm and anti-skin fungi.

# 35 / 双荚决明

*Senna bicapsularis*

豆科 Leguminosae　决明属 *Senna*

## 植物形态

直立灌木。多分枝，无毛。叶长7~12cm，有小叶3~4对；叶柄长2~4cm；小叶倒卵形或倒卵状长圆形，膜质，长2~4cm，宽约2cm，顶端圆钝，基部渐狭，偏斜，下面粉绿色，侧脉纤细，在近边缘处呈网结。总状花序生于枝条顶端的叶腋间，常集成伞房花序状，长度约与叶相等，花鲜黄色，直径约2cm；雄蕊10枚，7枚能育，3枚退化而无花药，能育雄蕊中有3枚特大，高出于花瓣，4枚较小，短于花瓣。荚果圆柱状，膜质，直或微曲，长10~18cm，直径1~2cm，缝线狭窄。种

## PLANT MORPHOLOGY

Upright shrub, much branched, glabrous. Leaves 7-12cm in length, with 3-4 pairs of lobules; petiole 2-4cm long; lobules obovate or obovate-oblong, membranous, 2-4cm long, 2cm wide, apex rounded and obtuse, base narrowing, skewed, pinkish-green underneath, slender side veins, with net knots near the edges. Racemes are born in the leaf axils at the top of the branches, often integrated into corymbs, about the same length as the leaves, bright yellow flowers, about 2cm in diameter; 10 stamens, 7 fertile, 3 degenerated without anthers, can Among the stamens, 3 are extra-large, taller than the petals, and 4 are smaller and shorter than the petals. The pods are cylindrical, membranous, straight or slightly curved, 10-18cm long, 1-2cm in diameter, with narrow sutures; two rows of seeds.

子二列。

## 生态分布

原产美洲热带地区，现广布于全世界热带地区。

## 生长习性

耐干旱，不抗风，不耐积水。对土壤水肥条件要求不严。

## 用途、价值

可作绿肥、绿篱及观赏植物。

## ECOLOGICAL DISTRIBUTION

Originally produced in tropical regions of America, it is now widely distributed in tropical regions all over the world.

## GROWTH HABIT

Resistant to drought, wind, or stagnant water. Lax requirements for soil water and fertilizer conditions.

## APPLICATION,VALUE

It can be used as green manure, hedgerow and ornamental plants.

# 36 / 喙荚云实

豆科 Leguminosae　云实属 Caesalpinia
*Caesalpinia minax*
别名：石莲子、老鸦枕头、鬼棒头

## 植物形态

　　攀缘灌木，有刺。各部被短柔毛。二回羽状复叶，长20~45cm；羽片5~8对，小叶6~12对，椭圆形或长圆形，长2~4cm，宽1~1.8cm。总状花序或圆锥花序顶生；苞片卵状披针形；萼片5；花瓣5，白色，有紫色斑点，倒卵形；花期4~5月。荚果，长圆形，长7~13cm，宽4~5cm，先端圆钝而有喙，果瓣表面密生针状刺，有种子4~8颗；果期7月。种子椭圆形，长16~19mm，宽9~11mm。

## 生态分布

　　分布于中国、巴基斯坦、越南、印

## PLANT MORPHOLOGY

　　Climbing shrub with thorns. Each part is pubescent. Secondary pinnate compound leaves, 20-45cm in length; 5-8 pairs of pinnae, 6-12 pairs of leaflets, elliptic or oblong, 2-4cm in length and 1-1.8cm in width. Racemes or panicles, terminal; cells ovate lanceolate; sepals 5; petals 5, white, with purple spots, obovate; flowering period from April to May. Pods oblong, 7-13cm in length and 4-5cm in width, with blunt and beaked apex, and dense needle like spines on the surface of fruit petals, with 4-8 seeds; the fruiting period July. Seeds elliptic, 16-19mm in length and 9-11mm in width.

## ECOLOGICAL DISTRIBUTION

　　Distributed in China, Pakistan, Vietnam, India, Indonesia, etc.

度、印度尼西亚等地。

## 生长习性

　　喜温暖、湿润的环境。生长于山沟、溪旁或灌丛中。

## 用途、价值

　　种子入药，名石莲子，无毒，有开胃进食、祛湿祛热的功效。果实美观，具有较高的观赏价值。

## GROWTH HABIT

Favoring warm and humid environment. It grows in gullies, streams or shrubs.

## APPLICATION,VALUE

The seed is used as medicine, famous stone lotus seed, non-toxic, has the effect of appetizer eating, dehumidification and heat removal. The fruit is beautiful and has high ornamental value.

## 37 / 金边丝兰

*Yucca aloifolia*
龙舌兰科 Agavaceae　丝兰属 *Yucca*
别名：金叶丝兰、金边凤尾兰、金边剑麻

### 植物形态

常绿灌木。茎短。叶基部簇生，呈螺旋状排列，叶片坚厚，长50~80cm，宽4~8cm，顶端具硬尖刺，叶面有皱纹，浓绿色而被少量白粉，老叶具少数丝状物。圆锥花序，花杯形，下垂，白色，外缘绿白色略带红晕；花轴发自叶丛间；花瓣匙形，6枚；花期6~11月，于夜间开放。蒴果长圆状卵形，长5~6cm。

### 生态分布

原产于北美洲。温暖地区广泛栽培。

### 生长习性

喜阳光充足、通风良好的环境，耐寒；对土壤适应性很强，适生于排水良好、肥沃的砂质土壤。

### 用途、价值

可盆栽观赏，适宜植于公园花坛、住宅旁等美化环境；对氨气、乙烯等都有一定的抗性，可用于工矿区绿化环境。

### PLANT MORPHOLOGY

Evergreen shrub. Stem short. Leaf base clustered and arranged in a spiral shape. Leaf blade firm and thick, 50-80cm in length and 4-8cm in width, with hard spines at the top. Leaf surface wrinkled, dark green and covered with a small amount of white powder. Elder leaves with a few filaments. Panicle, cup-shaped, drooping, white, with green and white outer edge, slightly reddish; rachis from leaves; petals spatulate; flowering period from June to November, blooming night. Capsule, oblong ovate, 5-6cm in length.

### ECOLOGICAL DISTRIBUTION

Native to North America. Widely cultivated in warm areas.

### GROWTH HABIT

Favoring sunny, well-ventilated environment; cold resistant. It has strong adaptability to soil, suitable for sandy soil with good drainage and fertility.

### APPLICATION,VALUE

It can be potted for viewing, suitable for planting in the garden, beside the residence, beautifying the environment; it has certain resistance to ammonia and ethylene, and can be used for greening environment in industrial and mining areas.

# 38 / 红花玉芙蓉

玄参科 Scrophulariaceae　玉芙蓉属 *Leucophyllum*

*Leucophylum frutescens*

## 植物形态

常绿小灌木。高1.5~2.5m。老枝灰褐色，嫩枝密被星状茸毛。叶互生，倒卵形，长1.2~2.5cm，质地厚，密被银白色茸毛，全缘。花单生，花萼裂片长椭圆状披针形；花冠紫红色，钟形，5裂，内部被毛；花期夏、秋两季。蒴果。

## 生态分布

原产中美洲、北美洲。分布于中国、巴基斯坦、美国及墨西哥等地。

## 生长习性

喜光，耐热、耐旱，不耐阴。适生

## PLANT MORPHOLOGY

Evergreen dwarf shrub. Plant height 1.5-2.5m. Elder branches grayish brown, and young branches densely covered with stellate villi. Leaves alternate, obovate, 1.2-2.5cm in length, thick texture, densely covered with silvery white tomentose, entire. flower solitary, the calyx lobes long elliptic lanceolate; the corolla purplish red, bell shaped, five lobed, and interior hairy; flowering period summer and autumn. Capsule.

## ECOLOGICAL DISTRIBUTION

Native to Central America and North America. Distributed in China, Pakistan, the United States and Mexico, etc.

温度23~32℃，可生于贫瘠的砂壤土中。

## 用途、价值

树形优美、花色艳丽，是优良的观赏植物，丛植于庭院、公园等；可用于海滨绿化。

## GROWTH HABIT

Favoring sunshine, heat resistance, drought tolerance, not shade tolerance. The suitable temperature for growth is 23-32°C, which can grow in poor sandy loam soil.

## APPLICATION,VALUE

It is an excellent ornamental plant with beautiful shape and gorgeous flowers. It can be planted in gardens and parks and can be used for seaside greening.

# 39 / 朱缨花

*Calliandra haematocephala*
豆科 Leguminosae　朱缨花属 *Calliandra*
别名：红合欢、红绒球、美蕊花、美洲合欢

## 植物形态

落叶灌木。高1~3m。小枝褐色。二回羽状复叶，羽片1对；小叶7~9对，斜披针形，长2~4cm，宽7~15mm；托叶卵状披针形，宿存。头状花序，腋生；花萼钟形，绿色；花冠淡紫红色，裂片反折；雄蕊凸露于花冠之外；花期8~9月。荚果，线状倒披针形，长6~11cm，宽5~13mm，暗棕色；果期10~11月。种子5~6枚，长圆形，长7~10mm，宽3~4mm，棕色。

## 生态分布

原产南美。分布于中国、印度、巴

## PLANT MORPHOLOGY

Deciduous shrub. Plant height 1-3m. Branchlets brown. Secondary pinnate compound leaves, 1 pair of pinnae; 7-9 pairs of leaflets, obliquely lanceolate, 2-4cm in length and 7-15mm in width; stipules ovate lanceolate, persistent. Inflorescence capitulum, axillary; calyx bell shaped, green; corolla purplish red, lobes reflexed; stamens protruding outside corolla; flowering period from August to September. Pods, linear oblanceolate, 6-11cm in length, 5-13mm in width, dark brown; fruiting period from October to November. Seeds 5-6, oblong, 7-10mm in length, 3-4mm in width, brown.

## ECOLOGICAL DISTRIBUTION

Native to South America. Distributed in China,

基斯坦、越南、埃及、巴西等地。

## 生长习性

喜光，喜温暖湿润气候，不耐寒；适生于深厚肥沃的酸性土壤。

## 用途、价值

花色艳丽，是优良的观花树种，适宜在园林绿地或庭院中栽植，亦可盆栽。树皮可药用，具有利尿、驱虫等功效。

India, Pakistan, Vietnam, Egypt, Brazil , etc.

## GROWTH HABIT

Light-favored, favoring warm and humid climate, unhardy; suitable for deep and fertile acid soil.

## APPLICATION, VALUE

Flower color gorgeous, fine flower tree species, suitable for planting in the garden green space or courtyard, also can be potted. Bark applicable in medicine, with diuretic, anthelmintic and other effects.

# 40 / 月季花

*Rosa chinensis*

蔷薇科 Rosaceae  蔷薇属 *Rosa*

别名：月月花、月月红、玫瑰

## 植物形态

直立灌木。高1~2m。小枝圆柱形，有短粗的钩状皮刺。小叶3~5对，宽卵形或卵状圆柱形，长2~6cm，宽1~3cm，具锐锯齿；托叶大部贴生于叶柄，仅顶端分离部分成耳状，有腺毛。花集生，稀单生；萼片卵形，边缘常有羽状裂片，稀全缘；花瓣重瓣或半重瓣，红色、粉红色或白色，倒卵形；花期4~9月。果卵球形或梨形，红色；果期6~11月。

## 生态分布

分布于中国、巴基斯坦等地。

## PLANT MORPHOLOGY

Erect shrub. Plant height 1-2m. Branchlets cylindrical, with short thick hook like prickles; leaflets 3-5 pairs, broadly ovate or ovate cylindrical, 2-6cm in length, 1-3cm in width, acutely serrate; stipules mostly attached to petioles, only the apical part separated into ears, with glandular hairs. Flowers aggregated and rarely solitary; sepals ovate, with pinnate lobes on the edge, rarely entire; petals double or semidouble, red, pink or white, obovate; flowering period from April to September. Fruit ovoid or pear shaped, red; fruiting period from June to November.

## ECOLOGICAL DISTRIBUTION

Distributed in China, Pakistan, etc.

## 生长习性

喜温暖、日照充足、空气流通的环境。适生温度15~28℃。适生于疏松肥沃、富含有机质、微酸性、排水良好的壤土。

## 用途、价值

可用于园林绿化，是主要的观赏花卉。可作花篱、花屏、花墙，有净化空气、降低噪声污染的作用；可吸收硫化氢、氟化氢等有害气体，对二氧化硫、二氧化氮等有较强的抵抗能力。花可提取香料，也可作花茶；根、叶、花均可入药，具有活血消肿、消炎解毒等功效。

## GROWTH HABIT

Favoring warm, sunny and airy environment. Suitable for 15-28°C temperature. It is suitable for loam with loose, fertile, rich in organic matter, slightly acidic and well drained.

## APPLICATION, VALUE

It can be used for landscaping and is the main ornamental flower. It can be used as flower fence, flower screen and flower wall, which can purify the air and reduce noise pollution. It can absorb harmful gases such as hydrogen sulfide and hydrogen fluoride, and has strong resistance to sulfur dioxide and nitrogen dioxide. Flowers can extract spices or make scented tea. Roots, leaves and flowers can all be used as medicine, and have the effects of promoting blood circulation, reducing swelling, diminishing inflammation and detoxifying.

# 41 / 沙漠玫瑰

*Adenium obesum*

夹竹桃科 Apocynaceae　沙漠玫瑰属 *Adenium*

别名：天宝花、阿拉伯沙漠玫瑰、索马里沙漠玫瑰

## 植物形态

多肉灌木或小乔木。高1~4m。根肥大肉质，树干膨大。叶互生，集生枝端，倒卵形至椭圆形，长达8~15cm，宽2~4cm，肉质，近无柄。总状花序，顶生，花冠漏斗状，5裂，外缘红色至粉红色，中部色浅，裂片边缘波状；花期5~12月。种子有白色柔毛。

## 生态分布

原产非洲。分布于中国、突尼斯、阿尔及利亚、摩洛哥、巴基斯坦等地。

## PLANT MORPHOLOGY

Succulent shrub or small tree. Plant height 1-4m. Root plump and fleshy, and trunk expanded. Leaves alternate, with aggregated branches, obovate to elliptic, 8-15cm in length and 2-4cm in width, fleshy, nearly sessile. Racemes, terminal, corolla funnel-shaped, 5-lobed, outer margin red to pink, middle light color, lobes edge undulate; flowering period from May to December. Seeds white pilose.

## ECOLOGICAL DISTRIBUTION

Native to Africa. Distributed in China, Tunisia, Algeria, Morocco, Pakistan, etc.

## GROWTH HABIT

Favoring high temperature, drought and sunny environment. Suitable for 25-30°C temperature. It

## 生长习性

喜高温、干旱、阳光充足的环境，适生温度25~30℃；不耐阴，忌涝，忌浓肥和生肥，畏寒冷。适生于富含钙质、疏松透气、排水良好的砂质壤土。

## 用途、价值

植株矮小，花鲜红艳丽，可布置小庭院，盆栽观赏和装饰。花可药用，内服治反胃、吐血等；乳汁毒性较强，误食会引起心跳加速、心律不齐。

cannot bear shade, avoid waterlogging, avoid thick fat and raw fat, fear cold. It is suitable for sandy loam with rich calcium, loose air permeability and good drainage.

### APPLICATION,VALUE

The plants are short and the flowers are bright red, which can be decorated in small courtyards and potted for viewing and decoration. Flowers can be used medicinally to treat nausea and vomiting blood. Milk is highly toxic, and eating it by mistake will lead to accelerated heartbeat and arrhythmia.

# 42 / 花叶青木

*Aucuba japonica* var. *variegata*
丝缨花科 Garryaceae　桃叶珊瑚属 *Aucuba*
别名：洒金珊瑚、洒金日本珊瑚、洒金桃叶珊瑚

## 植物形态

常绿灌木。高1~3m。树皮幼时绿色，后转为灰绿色。叶对生，肉革质，矩圆形，长8~20cm，宽5~12cm，绿色，缘疏生粗齿牙，叶面具黄斑。圆锥花序顶生；花瓣近卵形或卵状披针形，暗紫色；花期3~4月。果卵圆形，长15~20mm，直径5~10mm，暗紫色或黑色。种子1枚。

## 生态分布

原产日本。热带地区有栽培。

## PLANT MORPHOLOGY

Evergreen shrub. Plant height 1-3m. Bark green when young, turning gray green later. Leaves opposite, fleshy and leathery, oblong, 8-20cm in length and 5-12cm in width, green, margin sparsely coarse teeth, leaf with macula. Panicle, terminal; petals subovate or ovate lanceolate, dark purple; flowering period from March to April. Fruit ovoid, 15-20mm in length, 5-10mm in diameter, dark purple or black. Seed 1.

## ECOLOGICAL DISTRIBUTION

Native to Japan. Cultivated in the tropics.

## GROWTH HABIT

Favoring warm and humid environment, and not cold-resistant. Suitable for temperature is 15-

## 生长习性

喜温暖、阴湿环境，不耐寒。适生温度15~25℃，适应性强，适生于疏松肥沃的弱酸性或中性壤土。

## 用途、价值

树形直立优美，枝繁叶茂，可盆栽，宜配植于门庭两侧，或在公园及庭园栽培作观赏植物。

25°C. Favored growing in loose and fertile weak acid or neutral loam.

## APPLICATION,VALUE

The tree is upright and beautiful, with luxuriant branches and leaves. It is suitable to be planted on both sides of the gate, and can be potted and cultivated as ornamental plants in parks and gardens.

# 43 / 基及树

## 植物形态

　　灌木。高1~3m。树皮褐色，分枝较多，幼枝疏被短硬毛。叶互生或簇生于短枝；革质；倒卵形或匙形；长1.5~3.5cm，宽1~2cm，先端骤尖或圆，基部渐窄楔形下延成短柄，上部叶缘具齿状，两面疏被短硬毛。聚伞花序，花冠钟状，白或稍红色。核果近球形，直径约5mm，内果皮骨质，具网状纹饰；花果期11月至翌年4月。种子4枚。

## 生态分布

　　分布于亚洲南部、东南部及大洋洲的巴布亚新几内亚及所罗门群岛等地。

## PLANT MORPHOLOGY

　　Shrub. Plant height 1-3m. Bark brown, with many branches, and young branches sparsely bristly. Leaves alternate or short branches; leaves leathery; obovate or spoon-shaped; 1.5-3.5cm in length, 1-2cm in width. Apex sharply pointed or round, and base narrowed and wedge-shaped and extended into a short stalk. Upper leaf margin toothed, and both sides sparsely covered with short bristles. Cymes, corolla campanulate, white or slightly red. Drupe nearly spherical, about 5mm in diameter, endocarp bony, with reticulate pattern, flowering and fruiting period from November to April of the following year. Seeds 4.

## ECOLOGICAL DISTRIBUTION

　　Distributed in South and Southeast Asia and

## 生长习性

喜温暖湿润气候，不耐寒，不耐阴，萌芽力强，耐修剪。适生于疏松肥沃、排水良好的微酸性土壤。

## 用途、价值

绿化观赏树种，适于制作盆景，也可配置庭园中观赏。生长力强，耐修剪，常种植作绿篱。

in Papua New Guinea and Solomon Islands in Oceania, etc.

### GROWTH HABIT

Favoring warm and humid climate, not cold and shade resistance, strong germination, and pruning-resistant. Favored growing in loose, fertile soil and slightly acidic soil with good drainage.

### APPLICATION, VALUE

Greening ornamental tree species, suitable for making bonsai, and arranged in gardens for viewing. Strong growth ability, resistance to pruning, often planted as a hedgerow.

第三章

# 草本

PART 3　HERB

# 01 / 细叶结缕草

禾本科 Poaceae
*Zoysia pacifica*
结缕草属 *Zoysia*
别名：天鹅绒草

## 植物形态

多年生草本。高4~15cm。具匍匐茎，秆纤细。叶鞘无毛，紧密裹茎；叶舌膜质，长约0.2cm，顶端碎裂为纤毛状；鞘口具丝状长毛，小穗窄狭，黄绿色或有时略带紫色，披针形，顶端及边缘膜质，具不明显的5脉。花果期8~12月。

## 生态分布

原产菲律宾。分布于中国、巴基斯坦、美国、英国、法国等地。

## PLANT MORPHOLOGY

Perennial herb. Plant height 4-15cm. With stolons, slender culms. Leaf sheath glabrous, tightly wrapped around the stem; leaf tongue membranous, about 0.2cm in length, and tip fragmented into cilia; leaf sheath with long filamentous hairs, narrow spikelets, yellow-green, or sometimes purple, lanceolate, top and edge membranous, with inconspicuous 5-vein. Flowering and fruiting period from August to December.

## ECOLOGICAL DISTRIBUTION

Native to the Philippines. Distributed in China, Pakistan, America, England, France, etc.

## GROWTH HABIT

It is suitable for growing in tropical and

## 生长习性

适生于热带、亚热带地区，适生温度20~30℃，耐寒能力差，在低温5℃时会停止生长。喜光不耐阴。喜肥沃土壤，适生于弱酸性到弱碱性土壤。

## 用途、价值

优等牧草，草质柔嫩，牛、马、羊均喜食；是铺建草坪的优良禾草，因草质柔软，尤宜铺建儿童公园。可用于花坛、草坪、运动场、机场等场所。

subtropical areas. The suitable temperature for growth is 20-30°C, and its cold tolerance is poor. It will stop growing at a low temperature of 5°C. Light-favored and not tolerant of shade. Likes fertile soil and suitable for weak acid to weak alkaline soil.

## APPLICATION,VALUE

Excellent forage grass, tender grass, cattle, horses and sheep are fond of eating. It is an excellent grass for paving lawns, and it is especially suitable for paving children's parks because of its soft quality. It can be used in flower beds, lawns, sports fields, airports and other places.

# 02 / 高粱

禾本科 Poaceae　高粱属 *Sorghum*
*Sorghum bicolor*
别名：蜀黍、荻粱、乌禾、木稷

## 植物形态

　　一年生草本。高1~3m。秆较粗壮，直立，基部节上具支撑根。叶线形至线状披针形，长30~70cm，宽3~10cm，基部圆或微呈耳形，背面淡绿色或有白粉，边缘软骨质，具细小刺毛，中脉较宽。圆锥花序疏松，总梗直立或微弯曲；主轴具纵棱，疏生细柔毛，分枝3~7枚；花果期6~9月。颖果，长3~5mm，宽2~4mm，淡红色至红棕色，顶端微外露。

## 生态分布

　　分布于中国、韩国、日本、巴基斯

## PLANT MORPHOLOGY

　　Annual herb. Plant height 1-3m. Culms stout and erect with supporting roots on the basal nodes. Leaves linear to linear-lanceolate, 30-70cm in length and 3-10cm in width, base round or slightly ear-shaped, the back light green or with white powder, margin cartilaginous, with minute spines, and midrib wide. Panicles loose, total peduncle erect or slightly curved; main shaft with longitudinal edge, sparsely pilose, with 3-7 branches; flowering and fruiting period from June to September. Caryopsis, 3-5mm in length, 2-4mm in width, light red to reddish brown, slightly exposed at the top.

## ECOLOGICAL DISTRIBUTION

　　Distributed in China, South Korea, Japan, Pakistan, etc.

坦等地。

## 生长习性

　　喜温，抗旱、耐涝、耐高温，适生温度20~30℃。喜光，是C4作物，全生育期都需要充足的光照。平原、干旱丘陵、瘠薄山区均可种植。

## 用途、价值

　　可食用，可制成面条、面卷、煎饼等；亦可制糖、酿酒和制酒精等。

## GROWTH HABIT

　　Thermophilic, barren, flood-enduring, heat-resistant, suitable temperature is 20-30°C. light-favored , is a C4 crop, and needs sufficient sunlight throughout its growth period. It can be planted in plains, arid hills and barren mountainous areas.

## APPLICATION,VALUE

　　Edible, can be made into noodles, rolls, pancakes, etc. It can also be used to make sugar, wine and alcohol.

# 03 / 玉米

*Zea mays*
禾本科 Poaceae　玉蜀黍属 *Zea*
别名：苞谷、苞米、棒子、玉蜀黍、珍珠米

## 植物形态

一年生高大草本。秆直立，高1~4m。基部各节具气生支柱根。叶片宽大，线状披针形，边缘呈波状皱褶，中脉粗壮。顶生雄性圆锥花序，被细柔毛；雌花序被多数鞘状苞片所包藏。颖果，球形或扁球形，花、果期7~9月。

## 生态分布

原产美洲地区。分布于中国、巴基斯坦、美国、巴西、墨西哥、印度和罗马尼亚等地。

## 生长习性

喜温作物，适宜较高的温度。发芽适生温度25~35℃，40℃以上停止发芽。为C4植物，具较强的光合能力，需充足的水分。根系发达，需良好的土壤通气条件。

## 用途、价值

营养价值较高，是优良的粮食作物。资源极为丰富、廉价且易获得，具有许多生物活性，如抗氧化、降血糖、提高免疫力等。是畜牧业、水产养殖业等重要饲料来源，也是食品、医疗卫生、轻工业、化工业等原料之一。

## PLANT MORPHOLOGY

Annual tall herb. Culms erect, plant height 1-4m. Each section at the base with aerial roots. Leaves large, linear-lanceolate, wavy edge creases, midvein thick. Terminal male panicles, pilose, the female inflorescences enclosed by many sheath-like bracts. Caryopsis, spherical or oblate, flowering and fruiting period from July to September.

## ECOLOGICAL DISTRIBUTION

Native to America. Distributed in China, Pakistan, the United States, Brazil, Mexico, India and Romania, etc.

## GROWTH HABIT

Favoring warm crops, suitable for higher temperatures. The optimum temperature for germination is 25-35°C, and the germination stops above 40°C. Maize is a C4 plant with strong photosynthetic capacity. Sufficient moisture is required. The root system is well developed and requires good soil aeration conditions.

## APPLICATION,VALUE

It has high nutritional value and is an excellent food crop. Resources are extremely rich, cheap and easy to obtain, and have many biological activities, such as anti-oxidation, lowering blood sugar and improving immunity. It is an important feed source for animal husbandry and aquaculture, and it is also one of the raw materials for food, medical and health care, light industry and chemical industry.

# 04 / 狗牙根

*Cynodon dactylon*

禾本科 Poaceae　狗牙根属 *Cynodon*

别名：百慕大草

## 植物形态

多年生低矮草本。叶鞘微具脊，具柔毛；叶舌有1轮纤毛；叶线形，长1~12cm，宽0.1~0.5cm，通常无毛。穗状花序；小穗灰绿色或带紫色；外稃舟形，具3脉，背部明显成脊，被柔毛；内稃与外稃近等长，具2脉。颖果，长圆柱形，花果期5~10月。

## 生态分布

分布于中国、日本、俄罗斯、韩国、巴基斯坦等地。

## PLANT MORPHOLOGY

Low perennial herb. Leaf sheath slightly ridged, pilose; ligule with a ring of cilia; leaf linear, 1-12cm in length and 0.1-0.5cm in width, usually glabrous. Spikes; spikelets gray-green or purple; lemma boat-shaped, with 3-veined, back obviously ridged, pilose; palea and lemma nearly as long, with 2-veined. Caryopsis, long cylindrical, flowering period from May to October.

## ECOLOGICAL DISTRIBUTION

Distributed in China, Japan, Russia, South Korea, Pakistan, etc.

## GROWTH HABIT

Favoring warm. The optimum growth temperature is 20-30°C. The root system is shallow,

## 生长习性

喜温植物。适生温度20~30℃。根系分布较浅,夏季干旱时应及时灌溉。适生于排水良好、养分充足的土壤。

## 用途、价值

蛋白质含量高,可作为家畜饲料,草食性鱼类的优质青饲料,也可调制干草和青贮料;全草可入药,具有清血解热功效。草质柔软,根茎繁殖力强,是一种良好的水土保持植物;适于作为人工草甸种植。

so it should be irrigated in time when it is dry in summer. Suitable for fertile soil with good drainage and sufficient nutrients.

### APPLICATION,VALUE

Protein is high in content, which can be used as livestock feed, high-quality green feed for herbivorous fish, and can also be used to prepare hay and silage; The whole herb can be used as medicine, and has the effects of clearing blood and relieving fever. The grass is soft and the rhizome has strong reproductive ability, so it is a good plant for soil and water conservation. Suitable for planting as artificial meadow.

# 05 / 马齿苋

*Portulaca oleracea*

马齿苋科 Portulacaceae　马齿苋属 *Portulaca*

别名：马苋、五行草、瓜子菜、马齿菜、蚂蚱菜

## 植物形态

一年生草本。茎淡绿色或带暗红色。叶互生或近对生，扁平肥厚，倒卵形，长1~3cm，先端圆钝或平截，有时微凹，上面暗绿色，下面淡绿或带暗红色。花期5~8月。蒴果，卵球形，果期6~9月。种子多而细小，黑褐色。

## 生态分布

原产中国。分布于日本、俄罗斯、韩国、巴基斯坦、英国、美国等地。

## 生长习性

喜湿、耐旱、耐涝。适生温度20~30℃。具向阳性。适宜在各种田地和坡地栽培，适生于中性和弱酸性的土壤。

## 用途、价值

全草供药用，有清热利湿、解毒消肿等作用，种子明目；茎可食用。还可作兽药和农药，也是优良的饲料。

## PLANT MORPHOLOGY

Annual herb. Stem pale green or dark red. Leaves alternate or nearly opposite, flat and hypertrophy, obovate, 1-3cm in length, blade apex rounded or flat, sometimes slightly concave, dark green above, light green or dark red below. Flowering period from May-August. Capsule, ovoid, fruiting period from June to September. Seeds numerous and small, dark brown.

## ECOLOGICAL DISTRIBUTION

Native to China. Distributed in Japan, Russia, South Korea, Pakistan, England, America, etc.

## GROWTH HABIT

Wet-loving, drought and flood-enduring. The most suitable growth temperature is 20-30°C. It has phototropism. It is suitable for cultivation in various fields and sloping fields, and suitable for neutral and weakly acidic soils.

## APPLICATION,VALUE

The whole plant is used for medicine, which has the functions of clearing away heat and promoting diuresis, detoxifying and reducing swelling, and improving the eyesight of seeds; The stems are edible. It can also be used as veterinary drugs and pesticides, and is also an excellent feed.

## 06 / 苋

### 植物形态

一年生草本。高80~150cm。茎绿色或红色，常分枝。叶片长4~10cm，绿色、红色、紫色、黄色或部分绿色夹杂其他颜色。穗状花序下垂，花簇腋生，球形，绿色或黄绿色，顶端有长芒尖，花期5~8月。胞果，卵状长圆形，长2~3mm，环状横裂，包在宿存花被片内；果期7~9月。种子近球形或倒卵形。

### 生态分布

原产印度。分布于中国、巴基斯坦、日本、越南、泰国等地。

### 生长习性

喜温暖气候，耐热力强，不耐寒冷；适生温度20~30℃，20℃以下植株生长缓慢，10℃以下种子发芽困难。在高温短日照条件下，易开花结籽。对土壤要求不严，适生于偏碱性土壤。

### 用途、价值

茎叶可食用；根、果实及全草可入药，有明目、祛寒热的功效；叶可供观赏。

### PLANT MORPHOLOGY

Annual herb. Plant height 80-150cm. Stems green or red, often branched. Leaves 4-10cm in length, green, red, purple, yellow, or part of green mixed with other colors. Spikes, drooping, axillary flower clusters, spherical, green or yellow-green, with long awn tips at the top, flowering period from May to August. Utricle, ovate-oblong, 2-3mm in length, annular transverse split, enclosed in persistent perianth; fruiting period from July to September. Seeds nearly spherical or obovate.

### ECOLOGICAL DISTRIBUTION

Native to India. Distributed in China, Pakistan, Japan, Vietnam, Thailand, etc.

### GROWTH HABIT

Favoring warm climate, heat-resistant, not cold-resistant; suitable for temperature is 20-30°C, plants grow slowly below 20°C, and it is difficult for seeds to germinate below 10°C. Under high temperature and short sunshine conditions, it is easy to bloom and set seeds. Lax requirements on the soil, suitable for growing in alkaline soil.

### APPLICATION, VALUE

The stems and leaves are edible; the roots, fruits and the whole plant can be used as medicines, which have the effects of improving eyesight, fend off a chill and fever; the leaves are available for viewing.

# 07 / 地肤

*Kochia scoparia*
藜科 Chenopodiaceae　地肤属 *Kochia*
别名：扫帚苗、扫帚菜、观音菜、孔雀松

## 植物形态

一年生草本。高50~100cm。茎直立，淡绿色或略带紫红色，具多棱，分枝稀疏。叶披针形或条状披针形，常有3条明显主脉，边缘有疏生锈色绢状缘毛。穗状圆锥花序；花期6~9月。胞果，扁球形，膜质；果期7~10月。种子卵形，黑褐色。

## 生态分布

分布于中国、韩国、日本、印度、巴基斯坦、意大利等地。

## PLANT MORPHOLOGY

Annual herb. Plant height 50-100cm. Stem erect, pale green or purplish red, multi-sided, sparsely branched. Leaves lanceolate or strip-lanceolate, often with 3 distinct main veins and sparsely rusty silk-like cilia on the edges. Spike panicles, flowering period from June to September. Utricle, oblate, membranous, fruiting period from July to October. Seed ovoid, dark brown.

## ECOLOGICAL DISTRIBUTION

Distributed in China, South Korea, Japan, India, Pakistan, Italy, etc.

## GROWTH HABIT

Light-favored and thermophilus plant, drought-resistance, saline-alkali-resistance, not resistant to

## 生长习性

喜温、喜光、耐干旱、不耐寒，适应性较强。对土壤要求不严，较耐盐碱，肥沃土壤生长旺盛。

## 用途、价值

嫩茎叶可作蔬菜供食用，富含胡萝卜素；种子含油，供食用及工业用；可药用，有清湿热、利尿等功效。叶可作饲料。整株可用于布置花篱、花境，或数株丛植于花坛中央；盆栽可点缀和装饰厅堂、会场。

cold, and strong adaptability. Not strict to the soil quality, more saline-alkali resistant, thriving in fertile soil.

### APPLICATION,VALUE

Tender stems and leaves can be used as vegetables for food, rich in carotene; seeds contain oil, for food and industrial use; medicinal, has the effects of clearing dampness and heat, diuresis and so on. The leaves can be used as feed. The whole plant can be used to arrange flower hedges, flower borders, or plant several plants in the center of flower beds; potted plants can be used to embellish and decorate halls and venues.

# 08 / 紫背万年青

*Tradescantia spathacea*

鸭跖草科 Commelinaceae 紫露草属 *Tradescantia*

别名：紫锦兰、紫兰、蚌花

## 植物形态

多年生草本。丛生，具节。高20~60cm。叶互生，半抱茎，稍肉质，长圆状披针形；叶片表面深绿色，背面紫色；叶鞘口部时有长柔毛；花朵簇生，花梗紫色；花冠蓝色，分离；花丝具羽状长须毛，花期8~10月。蒴果，常不育。

## 生态分布

原产墨西哥和西印度群岛。分布于中国、印度、巴基斯坦等地。

## PLANT MORPHOLOGY

Perennial herb. Rosette, with node. Plant height 20-60cm. Leaves intergrowth, half-stem-clasping, slightly fleshy, oblong-lanceolate; the surface of the leaf dark green, and back purple; the leaf sheath pilose at the mouth; the flowers clustered and the peduncle purple; corolla blue, separated; the filaments pinnately long whisker, flowering period from August to October. Capsule, often infertile.

## ECOLOGICAL DISTRIBUTION

Native to Mexico and the West Indies. Distributed in China, India, Pakistan, etc.

## GROWTH HABIT

Favoring warm plant, shade-resistant. Suitable for temperature is 15-25°C, and the temperature in

## 生长习性

喜温植物。喜光耐阴，适生温度15~25℃，冬季温度不能低于5℃。适生于疏松肥沃、排水良好的砂壤土。

## 用途、价值

叶面光亮翠绿，叶背深紫，是常见的盆栽观叶植物，具有较高的观赏价值，亦可吸附有害气体，具有净化空气的作用。

winter cannot be lower than 5°C. Favored growing in sandy soil, loose and fertile soil with good drainage.

### APPLICATION,VALUE

The leaf surface is bright and emerald green, and the back of the leaf is deep purple. It is a common potted foliage plant with high ornamental value. It can also absorb harmful gases and purify the air.

# 09 / 锦绣苋

## 植物形态

多年生草本。高20~50cm。茎直立或基部匍匐，上部四棱形，在顶端及节部有贴生柔毛。叶片顶端有凸尖，边缘皱波状，呈绿色或红色，杂以红色或黄色斑纹。头状花序，无总花梗，丛生；花被片卵状矩圆形，白色，长0.3~0.4cm，凹形；花期8~9月。果实不发育。

## 生态分布

原产巴西。分布于中国、越南、缅甸、泰国、老挝、印度、巴基斯坦等地。

## 生长习性

喜高温，喜光。适生温度20~35℃，极不耐寒。适生于排水良好的土壤。

## 用途、价值

可作蔬菜食用。全株可入药，有清热解毒、凉血止血等功效。植株多矮小，叶色鲜艳，枝叶茂密，耐修剪，适合盆栽观赏。

## PLANT MORPHOLOGY

Perennial herb. Plant height 20-50cm. Stem erect or creeping, upper part quadrangular, with adnate pilose at the top and node. Top of the leaf having a convex tip, edge wrinkled, green or red, mixed with red or yellow markings. Inflorescence head, without peduncle, tufted; perianth pieces ovate-shaped oblong round, white, 3-4mm in length, concave, florescence from August to September. Fruit agenesis.

## ECOLOGICAL DISTRIBUTION

Native to Brazil. Distributed in China, Vietnam, Myanmar, Thailand, Laos, India, Pakistan, etc.

## GROWTH HABIT

Light-favored and heat-favored. Suitable for temperature is 20-35°C, and extremely not cold-resistant. Thriving in deep, loose and fertile soil.

## APPLICATION,VALUE

Edible as vegetables. Whole plant can be used as medicine, and has the effects of clearing away heat, detoxifying, cooling blood and stopping bleeding. Plants are short, with bright leaves, dense branches and leaves, pruning resistant, suitable for potted plants.

# 10 / 马唐

*Digitaria sanguinalis*
禾本科 Poaceae 马唐属 *Digitaria*

## 植物形态

一年生草本。高10~80cm。秆直立或下部倾斜，直径2~3mm。叶线状披针形，长5~15cm，宽0.4~1.5cm，基部圆形，边缘粗糙，具柔毛或无毛；叶鞘短于节间。总状花序；小穗椭圆状披针形；第一颖小，短三角形，无脉；第二颖具3脉，披针形，脉间及边缘大多具柔毛；第一外稃中脉平滑，边脉具小刺；第二外稃近革质，灰绿色。颖果；花果期6~9月。

## 生态分布

原产中国。分布于温带和亚热带山地。

## 生长习性

喜光照，喜湿，繁殖力强，植株生长快，分枝多。对土壤要求不严。

## 用途、价值

茎秆纤细，叶片柔软，是优良的饲草；可压制绿肥；全草可药用，有明目润肺的功效；亦可作固土、绿化等地被植物。

## PLANT MORPHOLOGY

Annual herb. Plant height 10-80cm. Stalk erector inclined at the bottom, 2-3mm in diameter. Leaves linear-lanceolate, 5-15cm in length, 0.4-1.5cm in width, base round, rough edges, pilose or glabrous; leaf sheaths shorter than internodes. Racemes; spikelets elliptic-lanceolate; first glume small, short triangular, without veins; the second glume having 3 veins, lanceolate, the veins and edges mostly pilose; the first lemma smooth, side veins having small spines; the second lemma nearly leathery, gray-green. Caryopsis; flowering and fruiting period from June to September.

## ECOLOGICAL DISTRIBUTION

Native to China. Distributed in temperate and subtropical mountains.

## GROWTH HABIT

Light-favored and humid-favored, strong fecundity, rapid plant growth and many branches. Not strict to the soil quality.

## APPLICATION,VALUE

Its slender stalks and soft leaves make it an excellent forage grass; it can suppress green manure; the whole grass can be used for medicinal purposes and has the effect of clear eyes moisturize the lungs; and used as ground cover plants such as soil consolidation and greening.

# 11 / 地毯草

禾本科 Poaceae　地毯草属 *Axonopus*
*Axonopus compressus*
别名：大叶油草

## 植物形态

多年生草本。高8~60cm。具长匍匐枝；节密生灰白色柔毛。叶鞘疏松，近鞘口处常疏生毛；叶片扁平，长5~10cm，宽0.5~1cm，近基部边缘疏生纤毛。花白色，总状花序，长4~8cm，呈指状排列在主轴上；小穗长圆状披针形，疏生柔毛，单生。颖果。

## 生态分布

原产热带美洲。世界各热带、亚热带地区有引种栽培。

## PLANT MORPHOLOGY

Perennial herb. Plant height 8-60cm. With long creeping branches; nodes densely grayish white pilose. Leaf sheath loose, and often sparse hairs near the leaf sheath opening; leaf flat, 5-10cm in length, 0.5-1cm in width, and sparse cilia on the margins near the base. Flowers white, racemes, 40-80mm in length, finger-like arranged on spindle; spikelets oblong-lanceolate, sparsely pilose, solitary; caryopsis.

## ECOLOGICAL DISTRIBUTION

Native to tropical America. Introduced and cultivated in tropical and subtropical regions of the world.

## 生长习性

喜潮湿的热带和亚热带气候，不耐旱，旱季休眠，不耐霜冻。适生于荫蔽及潮湿的砂土。

## 用途、价值

草质柔嫩，可作饲料，家畜、鱼喜食。匍匐枝蔓延迅速，每节都可生根和抽出新植株，植物体平铺地面成毯状，故称地毯草。根有固土作用，是铺建草坪、保持水土的优良草种。

## GROWTH HABIT

Favoring humid tropical and subtropical climate, not frost and not drought resistance, dormancy in dry season. Suitable for shade and wet sandy soil conditions.

## APPLICATION, VALUE

The grass is tender and can be used as fodder for livestock and fish. The creeping branches spread rapidly, and each node can take root and extract new plants. The plant body is flattened into a blanket, so called carpet grass. The roots have a soil-fixing effect, and are excellent grass species for paving and retaining soil.

# 12 / 叶下珠

*Phyllanthus urinaria*
叶下珠科 Phyllanthaceae　叶下珠属 *Phyllanthus*
别名：珠仔草、假油甘、龙珠草

## 植物形态

一年生草本。高10~60cm。茎基部多分枝；枝具翅状纵棱，被纵列疏短柔毛。叶长圆形或倒卵形，背面灰绿色，长4~10mm，近边缘具短粗毛；托叶卵状披针形。花2~4朵簇生叶腋；花期4~6月。蒴果，球形，直径1~2mm，红色，具小凸刺；果期7~11月。种子长1.2mm，橙黄色。

## 生态分布

原产中国。分布于印度、斯里兰卡、巴基斯坦等地。

## PLANT MORPHOLOGY

Annual herb. Plant height 10-60cm. Stem base multi-branched; branches with wing-like longitudinal ribs, sparsely puberulent by columns. Leaves oblong or obovate, back gray-green, 4-10mm in length, with short shag near the margin; stipules ovate-lanceolate. Flowers 2-4 clustered leaf axils, and the flowering period from April to June. Capsule, spherical, 1-2mm in diameter, red, with small convex thorns, fruiting period from July to November. Seeds 1.2mm in length, orange-yellow.

## ECOLOGICAL DISTRIBUTION

Native to China. Distributed in India, Sri Lanka, Pakistan, etc.

## 生长习性

喜温暖湿润气候，喜光，稍耐阴。适宜生长于弱酸性的砂质土壤中。

## 用途、价值

全株可入药，有清热利尿、明目、消积等功效；外用可消肿止痛。

## GROWTH HABIT

Like warm and humid climate, light-favored, slightly shade-tolerant. Suitable for growing in sandy soil.

## APPLICATION,VALUE

Applicable in medicine, effective in clearing heat and diuresis, improving eyesight, and eliminating accumulation; external use effective in reducing swelling and pain.

# 13 / 盐地碱蓬

*Suaeda salsa*

苋科 Amaranthaceae  碱蓬属 *Suaeda*

别名：黄须菜、翅碱蓬

## 植物形态

一年生草本。高10~80cm，全株呈绿色或紫红色。茎黄褐色。叶条形，半圆柱状，通常长1~3cm，枝上部的叶较短。花腋生，聚伞花序，在分枝上排列成有间断的穗状花序；花被半球形，稍肉质，具膜质边缘。胞果，成熟时常开裂。种子双凸镜形或歪卵形，直径2mm，黑色，具不清晰网点纹饰。花果期7~10月。

## 生态分布

原产中国。分布于欧洲及亚洲等地。

## PLANT MORPHOLOGY

Annual herb. Plant height 10-80cm, whole plant green or purplish red. Stems yellowish brown. Leaves strip-shaped, semi-cylindrical, usually 1-3cm in length, upper branch leaves shorter. Inflorescence cymose, arranged in discontinuous spikes on branches; perianth hemispherical, slightly fleshy, membranous margin. Utricle, often dehiscent at maturity. Seeds biconvex or skew-ovate, 2mm in diameter, black, with unclear dot ornamentation. Flowering and fruiting period from July to October.

## ECOLOGICAL DISTRIBUTION

Native to China. Distributed in Europe and Asia, etc.

## 生长习性

耐热、耐盐碱、耐干旱，生于盐碱土，常在海滩及湖边形成群落。

## 用途、价值

幼苗可作蔬菜食用；种子可榨油，是一种优质的蔬菜和油料作物。作为饲料，有助于提高家畜免疫力。富含钾盐，可用于化工、玻璃、制药等。盐地碱蓬具有改良土壤及降低土壤中重金属镉含量的作用。目前，以盐地碱蓬植物为主，结合其他生态修复的技术已大面积应用于生态治理。

## GROWTH HABIT

Heat, salt-alkali and drought-resistant, born in saline-alkali soil, often forms communities on beaches and lakeshores.

## APPLICATION,VALUE

Seedlings as vegetables; seeds squeezed into oil, a high-quality vegetable and oil crop. Making into feed to help improve the immunity of livestock. Rich in potassium salts and used in chemicals, glass, pharmaceuticals, etc. Effective in improving the soil and reducing the heavy metal cadmium in the soil. Mainly used in combination with other ecological restoration technologies and ecological management.

# 14 / 紫鸭跖草

*Tradescantia pallida*

鸭跖草科Commelinaceae　紫露草属 *Tradescantia*

别名：紫竹梅、紫竹兰、紫锦草

## 植物形态

多年生草本。高20~50cm。茎多分枝，带肉质，紫红色，下部匍匐状，节上常生须根。叶互生，基部抱茎而成鞘。花密生于花序柄上；花瓣蓝紫色，广卵形；花期7~9月。蒴果，椭圆形，有3条隆起棱线；果期9~10月。种子呈三棱状半圆形，淡棕色。

## 生态分布

原产墨西哥。分布于中国、越南、缅甸、泰国、老挝、印度、巴基斯坦等地。

## PLANT MORPHOLOGY

Perennial herb. Plant height 20-50cm. Stems multi-branched, fleshy, purple-red, prostrate at the lower part, nodes often having fibrous roots. Leaves alternate, and base holding the stem, turning sheath. Flowers densely born on the stalk of the inflorescence; petals blue-purple, wide oval; flowering period from July to September. Capsule, oval, three ridges, fruiting period from September to October. Seeds triangularly semicircular, light brown.

## ECOLOGICAL DISTRIBUTION

Native to Mexico. Distributed in China, Vietnam, Myanmar, Thailand, Laos, India, Pakistan, etc.

## 生长习性

喜温，不耐寒，忌阳光暴晒，喜半阴，对干旱有较强的适应能力，适宜肥沃、湿润的壤土。

## 用途、价值

全株可入药，是一种用途广泛的中草药；可提取天然色素，颜色鲜艳，具有无毒、无污染、绿色环保等优势；整个植株全年呈紫红色，特色鲜明，具有较高的观赏价值。

## GROWTH HABIT

Thermophilic, unhardy, avoids sun exposure and prefers half shade. Strong adaptability to drought. Suitable for fertile and humid soil.

## APPLICATION,VALUE

Applicable in medicine. A kind of Chinese herbal medicine with a wide range of uses; extractable natural pigment, brightly-coloured, with the advantages of non-toxic, pollution-free, and environmental protection; purple-red throughout the year, with distinctive characteristics, with high ornamental value.

# 15 / 蓖麻

## 植物形态

　　草质灌木。高0.5~5m。小枝、叶和花序通常被白霜，茎多液汁。叶互生，近圆形，长和宽40~80cm，掌状7~11裂，具锯齿；叶柄粗，长达40cm，中空，盾状着生，顶端具2盘状腺体，基部具腺体。总状花序或圆锥花序；花萼裂片卵状三角形；花期6~9月。蒴果卵球形或近球形，长1~5cm；果皮具软刺或平滑。种子椭圆形，长8~18mm，光滑，具淡褐色或灰白色斑纹；胚乳肉质。

## PLANT MORPHOLOGY

　　Herbaceous shrub. Plant height 0.5-5m. Branchlets, leaves, and inflorescences usually have frost, and stems have sap. Leaves alternate, suborbicular, 40-80cm in length and width, palmately 7-11-lobed, serrate; petiole thick, up to 40cm, hollow, peltate, with 2 discoid glands at the top and glands at the base. Raceme or panicle; calyx lobes ovate triangular; flowering period from June to September. Capsule ovoid or subglobose, 1-5cm in length; pericarp spiny or smooth. Seeds elliptic, 8-18mm in length, smooth, with light brown or gray white stripes; endosperm fleshy.

## ECOLOGICAL DISTRIBUTION

　　Native to Egypt, Ethiopia and India. Widely distributed in tropical areas.

## 生态分布

原产埃及、埃塞俄比亚和印度。现广布于热带地区。

## 生长习性

喜高温，不耐霜，酸碱适应性强。

## 用途、价值

种子可榨油，是重要工业用油；可制表面活性剂、干性油、增塑剂及弹性橡胶等，是高级润滑油原料，为化工、轻工、冶金、机电等工业和医药的重要原料。可用于园林绿化。叶、根可作药用，有缓泻作用，具毒，勿食用；经高温脱毒后可作饲料；茎皮富含纤维，为造纸和人造棉原料。

## GROWTH HABIT

Favoring high temperature, not frost resistant, and strong acid-base adaptability.

## APPLICATION, VALUE

Seeds can be used as oil, which is an important industrial oil; it can be used to make surfactants, drying oil, plasticizers and elastic rubber, etc. It is a high-grade lubricating oil raw material and an important raw material for chemical industry, light industry, metallurgy, electromechanical industry and medicine. It can be used for landscaping. Leaves and roots can be used as medicine, have laxative effect, are toxic, and should not be eaten; can be used as feed after high temperature detoxification; stem bark is rich in fiber, which is the raw material for papermaking and rayon.

# 16 / 李氏禾

*Leersia hexandra*
禾本科 Poaceae　假稻属 *Leersia*
别名：秕壳草

## 植物形态

多年生草本。节部膨大且密被倒生微毛。叶片长5~12cm，有时卷折；基部两侧下延与叶鞘边缘相愈合成鞘边。圆锥花序，长5~10cm，具角棱；小穗长3~5mm，颖不存在。颖果，长约4mm；花果期6~8月。

## 生态分布

原产中国。分布于中国、越南、缅甸、泰国、老挝、印度、巴基斯坦等地。

## 生长习性

耐高温，喜温暖，光照充足即可。吸水性好，抗涝，多生于河、湖、沼泽等地。对土壤要求不高。

## 用途、价值

可作为生态恢复工程中的先锋植物予以配置应用；对污水的适应能力较好，作为氧化塘处理植物应用广泛，效果明显。

## PLANT MORPHOLOGY

Perennial herb. Nodes swollen and densely inverted hairy. Leaves 5-12cm in length, sometimes curled up; both sides of the base extending downward and joint with the edge of the sheath to form a sheath edge. Inflorescence panicles, 5-10cm in length, angular; spikelets 3-5mm in length, glume absent. Caryopsis, about 4mm in length, flowering period from June to August.

## ECOLOGICAL DISTRIBUTION

Native to China. Distributed in China, Vietnam, Myanmar, Thailand, Laos, India, Pakistan, etc.

## GROWTH HABIT

Heat-resistant, thermophilic, and enough light. Good water absorption, flood-enduring. Born in rivers, lakes, swamps and other places. not high to the soil quality.

## APPLICATION,VALUE

Configured and applied as a pioneer plant variety in ecological restoration projects; good adaptability to sewage and widely used as an oxidation pond treatment plant with obvious effects.

# 17 / 风车草

*Cyperus involucratus*
莎草科 Cyperaceae　莎草属 *Cyperus*
别名：紫苏、旱伞草

## 植物形态

多年生草本。高30~150cm。茎近圆柱状，基部包裹无叶的鞘，鞘棕色。叶顶生为伞状。叶状苞片20枚，长近相等，向四周展开，平展，呈伞状；小穗轴不具翅；鳞片紧密覆瓦状排列，苍白色，具锈色或黄褐色斑点。小坚果，椭圆形或近三棱形，褐色；花果期8~11月。

## 生态分布

原产非洲，分布于中国、巴基斯坦等地。

## PLANT MORPHOLOGY

Perennial herb. Plant height 30-150cm. Stem nearly cylindrical, sheath at the base is leafless, brown. Leaves apex, umbrella-shaped. 20 bracts, approximately equal in length, spread out to all sides, flat, and umbrella-shaped; spikelet axis not winged; scales are arranged in dense complex tile shape, pale with rusty or yellowish brown spots. Small nuts, oval, nearly triangular, brown. Flowering and fruiting period from August to November.

## ECOLOGICAL DISTRIBUTION

Native to Africa. Distributed in China, Pakistan, etc.

## 生长习性

喜温、喜湿、不耐寒，适生温度15~25℃。适应性强，以保水性强的肥沃土壤最适宜。

## 用途、价值

是园林水体造景常用的观叶植物，也是室内观叶植物，除盆栽观赏外，也可用于制作盆景、水培或作插花材料。

## GROWTH HABIT

Thermophilic, wet-loving, unhardy. Suitable for 15-25°C temperature. Adaptable and suitable for growing in fertile soil with strong water retention.

## APPLICATION,VALUE

Leaf-watching plants commonly used in garden water landscaping, also an indoor foliage plant. In addition to potted plants, used for making bonsai materials, hydroponics or flower arrangement materials.

# 18 / 菟丝子

*Cuscuta chinensis*
旋花科 Convolvulacea 菟丝子属 *Cuscuta*
别名：无根藤、无叶藤、黄丝藤、鸡血藤、金丝藤、无根草

## 植物形态

一年生寄生草本。茎缠绕，黄色，纤细，直径约1mm。花序侧生，花簇生成小伞形或小团伞花序；花冠白色，壶形，长约3mm，裂片三角状卵形，顶端锐尖或钝，向外反折，宿存；鳞片长圆形，边缘长流苏状。蒴果，球形，直径约3mm，几乎全为宿存的花冠所包围，成熟时整齐周裂。种子2~50枚，淡褐色，卵形，长约1mm，表面粗糙。

## 生态分布

分布于中国各地、伊朗、阿富汗向东至日本、朝鲜、马达加斯加、澳大利

## PLANT MORPHOLOGY

Annual parasitic herb. Stems twined, yellow, slender, about 1mm in diameter. Inflorescences laterally, flower clusters form small umbels or small clusters; corolla white, pot shape, about 3mm in length, lobes triangular-ovate, apex acute or obtuse, reflexed outwards, persistent; scales oblong, with long fringe-shaped edges. Capsule, globose, about 3mm in diameter, almost completely surrounded by persistent corolla, neatly lobed when mature. Seeds 2-50, light brown, ovoid, about 1mm in length, rough surface.

## ECOLOGICAL DISTRIBUTION

Distributed in China, Iran and Afghanistan eastward to Japan, North Korea, Madagascar, Australia, etc.

亚等地亦有分布。

## 生长习性

喜高温湿润气候，对土壤要求不严，适应性较强。野生菟丝子常见于平原、荒地以及豆科、菊科、蒺藜科等植物地内，最喜寄生于豆科植物上。

## 用途、价值

是一味补肾、肝、脾的良药，有补肝肾、益精壮阳、止泻等功效。但其为大豆产区的有害杂草，并对胡麻、花生、马铃薯等农作物也有危害。

## GROWTH HABIT

Favoring warm and humid climate, not strict to the soil quality, strong adaptability. Commonly found in plains, wastelands, and Fabaceae, Asteraceae, Zygophyllaceae, and other plant areas, and most like to be parasitic on legumes.

## APPLICATION, VALUE

A good medicine for tonifying kidney, liver and spleen, effective in tonifying liver and kidney, supplementing essence and strengthening yang, and stopping diarrhea. Harmful weed in soybean production areas, and harmful to crops such as flax, peanuts and potatoes.

## 19 / 非洲天门冬

*Asparagus densiflorus*

天门冬科 Asparagaceae　天门冬属 *Asparagus*

别名：万年青、天冬、武竹、蓬莱松、密叶天门冬

## 植物形态

　　常绿草本。可攀缘，高0.3~1m。茎和分枝有纵棱。叶状枝每3枚成簇，扁平，条形，长1~3cm，宽0.2~0.5cm。茎上的鳞片状叶，基部具长0.3~0.5cm的硬刺，分枝上无刺。花白色，总状花序单生或成对；花被片矩圆状卵形，长约2mm；花期6~8月。浆果，直径5~10mm，熟时红色，具1~2枚种子；果期10~12月。

## 生态分布

　　原产非洲南部，现已被广泛栽培。

## PLANT MORPHOLOGY

　　Perennial herb. Climbable, plant height 0.3-1m. Stems and branches with longitudinal prisms. Leaf-like branches clustered every 3, flattened, bar-shaped, 1-3cm in length and 0.2-0.5cm in width. Scaly leaves on the stem with hard spines about 0.3-0.5cm in length at the base, no spines on branches. Flowers white, inflorescences solitary or paired; tepals rounded ovate, about 2mm in length; flowering period from June to August. Berry, 5-10mm in diameter, red when ripe, with 1-2 seeds, fruiting period from October to December.

## ECOLOGICAL DISTRIBUTION

　　Native to southern Africa and now widely cultivated.

## 生长习性

喜温植物，不耐寒，畏强光暴晒和高温，忌干旱和积水，适生于疏松肥沃、排水良好的砂壤土。

## 用途、价值

可供观赏；叶丛球状似松，适于艺术插花、盆栽观赏。

## GROWTH HABIT

It is a thermophilic plant, unhardy, afraid of strong light exposure and high temperature, avoid drought and stagnant water, suitable for loose and fertile, well-drained sandy loam soil.

## APPLICATION,VALUE

Available for viewing; the leaves are ball-shaped like pine, suitable for artistic flower arrangement and potted viewing.

# 20 / 芭蕉

## 植物形态

多年生草本。高3~4m。叶片长圆形，长2~3m，叶面鲜绿色，有光泽；叶柄粗壮，长30cm。花序顶生，下垂；苞片红褐色或紫色。浆果，三棱状，长圆形，长50~70mm，具3~5棱，近无柄，肉质，内具多数种子。种子黑色，宽5~10mm。

## 生态分布

原产于琉球群岛，分布于中国、巴基斯坦等地。

## 生长习性

喜温，喜湿，稍耐阴，不耐寒。茎分生能力强。忌土壤持续积水。在土层深厚、疏松肥沃和排水好的土壤中生长良好。

## 用途、价值

叶纤维为芭蕉布的原料，亦为造纸原料；可药用，假茎解热，叶利尿，花可治疗脑出血。园林中应用广泛，常与其他植物搭配种植，组合成景，也可作盆景。

## PLANT MORPHOLOGY

Perennial herb. Plant height 3-4m. Leaves oblong, 2-3m in length, leaves bright green and shiny; petioles stout, 30cm in length. Inflorescences apical, drooping; bracts reddish-brown or purple. Berry, triangular, oblong, 50-70mm in length, 3-5 edged, nearly sessile, fleshy, with most seeds. Seeds black, 5-10mm in width.

## ECOLOGICAL DISTRIBUTION

Native to the Ryukyu Islands. Distributed in China, Pakistan, etc.

## GROWTH HABIT

Thermophilic, wet-loving and slightly shade-tolerant, unhardy. Strong stem meristem. Avoid continuous accumulation of water in the soil. It grows well in deep, loose, fertile and well-drained soils.

## APPLICATION,VALUE

The leaf fiber is the raw material for banana cloth and papermaking; it can be used medicinally, the pseudostem is antipyretic, the leaves are diuretic, and the flowers can treat cerebral hemorrhage. It is widely used in gardens and often planted in combination with other plants, combined to form a landscape, or as a bonsai.

# 21 / 田菁

## 植物形态

一年生草本。高3~4m。茎绿色或带褐红色，微被白粉；基部有多数不定根，幼枝疏被白色绢毛，折断有白色黏液。羽状复叶，叶轴具沟槽；小叶20~30对，线状长圆形，两面被紫色小腺点。总状花序，疏被绢毛；花冠黄色，旗瓣横椭圆形至近圆形，先端时微凹，外面散生大小不等的紫黑点或线。荚果，细长，微弯，具黑褐色斑纹，果颈开裂；种子绿褐色，种脐圆形。花果期7~12月。

## PLANT MORPHOLOGY

Annual herb. Plant height 3-4m. Stem green, or brownish red, slightly covered with white powder; adventitious roots at the base, young branches sparsely covered with white silk hair, white mucus after breaking. Pinnate compound leaves ,blade shaft with groove; 20-30 pairs of lobules, linear oblong, with small purple gland spots on both sides. raceme, sliced seriate; corolla yellow, standard horizontally elliptical to nearly round, and apex slightly concave, with purple black dots or lines of varying sizes scattered on the outside. Pods slender, slightly curved, with dark brown markings, and neck cracked; seeds green-brown and the hilum round. Flowering and fruiting period from July to December.

## 生态分布

分布于中国、巴基斯坦等地。

## 生长习性

喜温、喜湿。春播土温达15℃时发芽，但出苗和苗期生长缓慢。适应性强、耐盐碱、耐涝、耐瘠薄、耐旱，抗病虫害及抗风能力强。

## 用途、价值

茎叶富含纤维素与粗蛋白，适用于制作成青贮或干草饲料，亦可发酵成绿肥。根系发达，生长速度快，具有防风固沙、水土保持及绿化环境的作用。

## ECOLOGICAL DISTRIBUTION

Distributed in China, Pakistan, etc.

## GROWTH HABIT

Thermophilic and wet-loving. Germination occurs when the soil temperature reaches 15°C in spring, but the seedling and seedling grow slowly. Strong adaptability, saline-alkali resistance, flood-enduring, barrenness, drought, disease and insect pests and wind resistance.

## APPLICATION,VALUE

The stems and leaves are rich in cellulose and crude protein, suitable for making silage or hay feed, and can also be fermented into green manure. The root system is developed and the growth rate is fast. It has the functions of windbreak and sand fixation, soil and water conservation and greening of the environment.

# 22 / 茅瓜

*Solena heterophylla*
葫芦科 Cucurbitaceae 茅瓜属 *Solena*
别名：牛奶子、山天瓜、老鼠冬瓜

## 植物形态

匍匐或攀缘草本。块根纺锤状，径粗1.5~2cm。茎、枝柔软，无毛，具沟纹。叶形变异极大，有时开裂，卵形或长圆形，裂片长圆状披针形、披针形或三角形，长8~12cm，宽1~5cm；叶面稍粗糙，背面灰绿色，叶脉凸起，基部心形。伞状花序，花极小，花梗纤细，长2mm，几无毛；花冠黄色，外面被短柔毛；花期5~8月。果长圆状或近球形，红褐色，长20~60mm，径20~50mm；果期8~11月。种子数枚，近圆球形或倒卵形，灰白色。

## 生态分布

分布于中国、巴基斯坦等地。

## 生长习性

常生长在海拔600~2600m的山坡路旁、林下、杂木林中或灌丛中。

## 用途、价值

可药用，含有一些酮、酸以及甾体等药用成分，具有清热解毒和化瘀散结等功效，入药后能治疗肿毒与烫伤、肺痈咳嗽以及水肿。

## PLANT MORPHOLOGY

Creeping or climbing herbs. Tubers spindle-shaped, 1.5-2cm in diameter. Stems and branches soft, glabrous, and grooved. Leaf shape highly variable, sometimes cracked, ovate or oblong. lobes oblong-lanceolate, lanceolate, or triangular, 8-12cm in length, 1-5cm in width; leaf surface slightly rough, the dorsal surface gray-green, veins convex, and base heart-shaped. Umbel-shaped inflorescence, extremely small flowers, peduncles slender, 2mm in length, almost glabrous; corolla yellow and puberulent outside, flowering period from May to August. Fruit oblong or nearly spherical, reddish-brown, 20-60mm in length, 20-50mm in diameter, and the fruiting period from August to November. Seeds several, nearly spherical or inverted ovate, greyish-white.

## ECOLOGICAL DISTRIBUTION

Distributed in China, Pakistan, etc.

## GROWTH HABIT

It often grows on the slopes of 600-2600m on the side of the road, under the forest, in the mixed forest or in the shrubland.

## APPLICATION,VALUE

It can be used as medicine, and contains some medicinal ingredients such as ketones, acids and steroids. It has the effects of clearing heat and detoxicating, and removing blood stasis and resolving masses. After being used as medicine, it can treat swelling and scald, lung carbuncle cough and edema.

# 23 / 佛甲草

## 植物形态

多年生草本。高10~20cm。叶轮生或对生，线形，长2~3cm，基部无柄。聚伞花序，顶生，疏生花，径4~8cm，中央有一朵花具短梗，另有分枝；花瓣5，黄色，披针形，长4~8mm，基部稍窄；花期4~5月。蓇葖果，长4~5mm；果期6~7月。种子小。

## 生态分布

原产中国。分布于巴基斯坦、日本等地。

## 生长习性

适应性强，耐高温，耐贫瘠，耐干旱，耐寒；严寒期地上部茎叶处于休眠期，翌年萌发新芽，早春即能覆盖地面。生长快，扩展能力强，根系纵横交错。

## 用途、价值

是优良的地被植物，也可盆栽观赏；与土壤结合紧密，有水土保持的作用，适宜作护坡草；用于屋顶绿化，可取代传统的隔热层和防水保护层。

## PLANT MORPHOLOGY

Perennial herb. Plant height 10-20cm. Leaves whorled or opposite, linear, 2-3cm in length, sessile at the base. Cymes, terminal, sparsely flowered, 4-8cm in diameter, flower with short stems and branches in the centre; petals 5, yellow, lanceolate, 4-8mm long, slightly narrow at base; flowering period from April to May. Follicles, 4-5mm long, fruiting period from June to July. Seed small.

## ECOLOGICAL DISTRIBUTION

Native to China. Distributed in Pakistan, Japan, etc.

## GROWTH HABIT

Heat-resistant. It is highly adaptable, resistant to barrenness, drought, cold-resistant. During the severe cold period, the aboveground stems and leaves are dormant, and new shoots germinate the following year, covering the ground in early spring. It grows fast, expands strongly, and crisscrosses the root system.

## APPLICATION,VALUE

It is an excellent ground cover plant that can also be potted for viewing; closely integrated with the soil and has the function of water and soil conservation, and is suitable for slope protection. It is used for the greening of the top of the building and can replace the traditional insulation layer and waterproof protection layer.

# 24 / 苘麻

*Abutilon theophrasti*

锦葵科 Malvaceae　苘麻属 *Abutilon*

别名：苘、车轮草、磨盘草、桐麻、白麻、青麻

## 植物形态

一年生草本。高1~2m。茎枝被柔毛。叶互生，圆心形，长5~10cm，基部心形，具细圆锯齿，两面密被星状柔毛。花单生叶腋；花梗被茸毛，近顶端具节；花萼杯状，密被茸毛；花黄色，花瓣倒卵形；顶端平截，轮状排列，密被软毛；花期7~8月。蒴果，半球形，直径约20mm，被粗毛，顶端具长芒。种子肾形，褐色，被星状柔毛。

## 生态分布

原产中国。分布于巴基斯坦、越南、印度、日本等地。

## PLANT MORPHOLOGY

Annual herb. Plant height 1-2m. Stems and branches pilose. Leaves alternate, heart-shaped, 5-10cm in length, heart-shaped at the base, with fine circular serrations, densely stellate pilose on both sides. Flower solitary leaf axils; peduncle pilose, close to apex nodular; calyx cup-shaped, densely tomentose; flower yellow, petals obovate; apex truncated, arranged in a wheel, densely villous; flowering period from July to August. Capsule, hemispherical, about 20mm in diameter, coarsely hairy, with long awns at the top. Seeds kidney-shaped, brown, and stellate pubescent.

## ECOLOGICAL DISTRIBUTION

Native to China. Distributed in Pakistan, Vietnam, India, Japan, etc.

## 生长习性

喜温暖湿润和阳光充足的环境，适生温度25~30℃，不耐寒，较耐旱，适生于疏松而肥沃的土壤。常见于路旁、荒地和田野间。

## 用途、价值

皮层纤维可供织麻布、搓绳索和加工成人造棉。全草可入药，有散风、清血热、活血等功效。

## GROWTH HABIT

Favoring warm and humid and sunny environment, grows at a suitable temperature of 25-30°C, is not cold-resistant, drought resistance, and is suitable for loose and fertile soil. It is commonly found on roadsides, in heaths and fields.

## APPLICATION,VALUE

The fibers of the skin can be used to weave linen, rub rope and process adult cotton. The whole herb can be used in medicine, and has the effects of dispersing wind, clearing blood heat, and activating blood.

# 25 / 大叶落地生根

*Kalanchoe daigremontiana*

景天科 Crassulaceae  伽蓝菜属 *Kalanchoe*

别名：宽叶落地生根、落地生根

## 植物形态

多年生肉质草本。高50~100cm。茎单生，直立，褐色，基部木质化。叶对生或轮生，下部叶片较大，常抱茎；叶片肉质，长三角形，具不规则褐紫色斑纹；叶缘有粗锯齿，锯齿处长出具有2~4片真叶的幼苗，碰触落地，易形成气生根，根深入土中即可生成新的植株。聚伞花序，顶生，多分枝，花钟形，红色，花瓣风干呈紫色，宿存不落；花期12月至翌年4月。

## 生态分布

产马达加斯加。分布于中国、巴基

## PLANT MORPHOLOGY

Perennial fleshy herbs. Plant height 50-100cm. Stems solitary, erect, brown, woody at base. Leaves opposite or whorled, the lower leaves larger, often amplexicaul; leaf blades fleshy, long triangular, with irregular brown-purple markings; leaf margin with thick serrations, seedlings with 2-4 true leaves growed from the serrations, easy to grow aerial roots after being touched and falling to the ground; they can grow into new plants when roots extend deeply in the soil. Cymes terminal, multi-branched, bell-shaped, red, petals purple after drying, persistent, blooming is from December to April of the following year.

## ECOLOGICAL DISTRIBUTION

Inative to Madagascar. Distributed in China,

斯坦等地。

## 生长习性

喜温暖和阳光充足的环境，适生温度13~19℃。不耐弱光，耐干旱；繁殖能力强，扦插易生根。适生于腐叶土、泥炭土和粗砂混合土。

## 用途、价值

具有绿化观赏价值，是优良的室内盆栽植物。

Pakistan, etc.

## GROWTH HABIT

It likes a warm and sunny environment, and the suitable temperature for growth is 13-19°C. It is not tolerant to low light, but drought-tolerant; it has strong reproduction ability, and is easy to take root after cuttage. It is suitable for growing in mulch soil, peat soil and coarse sand mixed soil.

## APPLICATION,VALUE

It has green ornamental value and is an excellent indoor potted plant.

# 26 / 罗勒

## 植物形态

一年生草本。具圆锥形主根及自其上生出的密集须根。茎直立，钝四棱形，上部微具槽，基部常染有红色，多分枝。叶卵圆形至卵圆状长圆形，长2.5~5cm，宽1~2.5cm，叶边缘具不规则牙齿或近于全缘；叶柄被微柔毛。总状花序顶生于茎、枝，均被微柔毛；苞片短于轮伞花序，边缘具纤毛，常具色泽；花梗明显，先端下弯；花冠淡紫色；花期7~9月。小坚果，卵珠形，黑褐色；果期9~12月。

## 生态分布

原产非洲、美洲及亚洲热带地区。分布于摩洛哥、阿尔及利亚、突尼斯、墨西哥、中国、巴基斯坦等地。

## 生长习性

喜温暖湿润气候，不耐寒，耐干旱，不耐涝，适生于排水良好、肥沃的砂质壤土或腐殖质壤土。

## 用途、价值

嫩叶可食，干燥花叶可泡茶饮，可作调料，有祛风、健胃、发汗的作用。全草可入药，有治疗头痛、打嗝胀气、粉刺等功效。

## PLANT MORPHOLOGY

Annual herbs. With a conical taproot and dense fibrous roots growing from it. Stems are erect, obtuse quadrangular, upper slightly grooved, base often red, much branched. Leaf blades ovate to ovate oblong, 2.5-5cm in length, 1-2.5cm in width, leaf margins irregularly toothed or nearly entire;

petiole pubescent. Racemes terminal in stems and branches, all puberulent; bracts shorter than verticillaster, margin ciliate, often colored; pedicel obvious, apex bent downward; corolla pale purple, florescence is from July to September. Nutlets ovoid, dark brown. Fruiting period from September to December.

## ECOLOGICAL DISTRIBUTION

Native to tropical regions of Africa, America and Asia. Distributed in Morocco, Algeria, Tunisia, Mexico, China, Pakistan, etc.

## GROWTH HABIT

It likes a warm and humid climate, and is not tolerant to cold, drought and waterlogging. It is suitable for growing in well-drained, fertile sandy loam or humus loam.

## APPLICATION,VALUE

The young leaves are edible, and the dried flowers and leaves can be used for tea, and can be used as a seasoning. It has the effects of expelling wind, invigorating stomach and sweating. The whole herb can be used as medicine to treat headaches, hiccups, flatulence, acne.

# 27 / 珍珠猪毛菜

*Salsola passerina*
苋科 Amaranthaceae　猪毛菜属 *Salsola*
别名：珍珠柴、雀猪毛菜

## 植物形态

高10~30cm。植株密生"丁"字毛，自基部分枝；小枝黄绿色，短枝缩短成球形。叶片锥形或三角形，长0.2~0.4cm，背面隆起，通常早落。花序穗状，生于枝条的上部；小苞片宽卵形，两侧边缘为膜质；花被片长卵形，背部近肉质，边缘为膜质；花被片在翅以上部分，向中央聚集成圆锥体；花期7~9月。种子横生或直立；果期8~9月。

## 生态分布

分布于中国、巴基斯坦等地。

## PLANT MORPHOLOGY

Subshrub. Plant height 10-30cm. Plant with dense T-shaped hairs, some branches extended from the base; the branchlets yellowish green, and short branches shortened into globose shape. Leaf blade conical or triangular, 0.2-0.4cm in length, abaxially convex, usually caducous. Inflorescence spicate, on the upper part of branches; bracteoles broadly ovate, with membranous margins on both sides; perianth segments long ovate, dorsal near fleshy, margin membranous; perianth segments above wing, gathered into a cone to the center; flowering period from July to September. Seeds transverse or erect, and fruiting period is from August to September.

## ECOLOGICAL DISTRIBUTION

Distributed in China, Pakistan, etc.

## 生长习性

旱生植物，抗风沙，耐寒，生于砾石质山坡、砾质滩地、沙砾质戈壁或黏土壤、盐碱湖盆地。

## 用途、价值

可改善荒漠及荒漠草原地带生态环境，水土保持，防风固沙。可作为家畜牧草，是重要的饲用植物。可种植于庭院、公园、道路。可药用，主治高血压。种子可作工业用油。

## GROWTH HABIT

Xerophyte, resistant to wind and sand and cold, grows on gravel hillside, gravel beach, sandy gobi or clay soil, saline alkali lake basin.

## APPLICATION, VALUE

It can improve the ecological environment of desert and desert steppe, protect soil and water, prevent wind and fix sand. It is an important forage plant. It can be planted in gardens, parks and roads. It can be used for medicine, mainly for hypertension. The seeds can be used as industrial oil.

# 28 / 钻叶紫菀

*Aster subulatus*
菊科 Asteraceae   紫菀属 *Aster*
别名：土柴胡、剪刀菜、燕尾菜、钻形紫菀

## 植物形态

　　一年生草本。高20~100cm。茎和分枝具粗棱，有时带紫红色。叶片披针状线形，先端锐尖或急尖，基部渐狭，叶无柄，基生叶在花期凋落。圆锥花序，具钻形苞叶；总苞钟形，绿色或先端带紫色，先端尖。瘦果，线状长圆形，长1~2mm，稍扁，疏被白色微毛；花果期6~10月。

## 生态分布

　　原产北美洲。分布在中国、巴基斯坦等地。

## PLANT MORPHOLOGY

　　Annual herb. Plant height 20-100cm. Stems and branches roughly ribbed, sometimes purplish red. Leaves lanceolate, apex acute or abruptly-acuminate, base attenuate, leaves sessile, basal leaves withered during the flowering period. Panicle with subulate bracts; involucres campanulate, apex green or purple, acute. Achene, linear oblong, 1-2mm in length, slightly flat, sparse white puberulous; flowering and fruit period from June to October.

## ECOLOGICAL DISTRIBUTION

　　Native to North America. Distributed in China, Pakistan, etc.

## 生长习性

喜湿润和阳光充足的环境，耐半阴，常见于山坡灌丛、草坡、沟边、路旁或荒地。

## 用途、价值

全草可入药，有清热解毒的功效，可治痈肿、湿疹等。

## GROWTH HABIT

It likes a warm, humid and sunny environment; it is tolerant to half shade, and is commonly found in hillside shrubs, grass slopes, ditch sides, roadsides or wastelands.

## APPLICATION,VALUE

The whole herb can be used as medicine, has the effects of clearing away heat and detoxification, and can cure carbuncle swelling, eczema, etc.

# 29 / 番茄

Lycopersicon esculentum
茄科 Solanaceae　番茄属 Lycopersicon
别名：洋柿子、番茄、西红柿

## 植物形态

一年生草本。高1~2m。植株被黏质腺毛。茎易倒伏，羽状复叶或羽状深裂，长10~40cm，小叶5~9对，卵形或长圆形，长5~7cm，基部楔形，偏斜，具不规则锯齿或缺裂；叶柄长2~5cm。花序梗长2~5cm，具3~7花；花梗长1~2cm；花萼辐状钟形，裂片披针形，宿存；花冠辐状，径2~3cm，黄色，裂片常反折。浆果，扁球形或近球形，肉质多汁液，橘黄或鲜红色。种子黄色，被柔毛。

## PLANT MORPHOLOGY

Annual herbs. Plant height 1-2m. Plants with densely viscid glandular hairs. Stems easy to lodging, pinnately compound leaves or pinnatipartite, 10-40cm in length, leaflets 5-9 pairs, ovoid or oblong, 5-7cm in length, base cuneate, oblique, with irregular serrations or dehiscence; petiole 2-5cm in length. Peduncle 2-5cm long, with 3-7 flowers; pedicel 1-2cm in length; calyx rotate campanulate, lobes lanceolate, persistent; corolla rotate, 2-3cm in diameter, yellow, lobes often reflexed. Berries, oblate or sub globose, succulent, orange or bright red. Seeds yellow and pilose.

## ECOLOGICAL DISTRIBUTION

Native to South America. Distributed in China, Pakistan, the United States, Argentina, Peru, Mexico, Spain, etc.

## 生态分布

原产南美洲。分布于中国、巴基斯坦、美国、阿根廷、秘鲁、墨西哥、西班牙等地。

## 生长习性

喜温、喜光，短日照植物，适生温度20~25℃。对土壤条件要求不严，适生于土层深厚、排水良好、富含有机质的肥沃壤土。

## 用途、价值

可食用。营养价值高，可加工成番茄酱、番茄汁等。含有"番茄素"，有抑制细菌降低胆固醇的作用，对高脂血症有益处；还有增加胃液酸度、健胃消食、清热解暑、保护心脏的作用。被称为神奇的菜中之果。

## GROWTH HABIT

It likes warmth, light, and is a short-day plant with an optimum temperature of 20-25°C. The soil conditions are not strict, and it is suitable for growing in fertile loam with deep soil layer, good drainage and rich organic matter.

## APPLICATION,VALUE

It is edible. It has high nutritional value and can be processed into tomato sauce, tomato juice, etc. It contains "tomatine", which has the effect of inhibiting bacteria, reducing cholesterol content, which is beneficial for hyperlipidemia. It can also increase the acidity of gastric juice, invigorate the stomach and promote digestion. It has the function of clearing heat and relieving heat and protecting the heart. It is known as the fruit of the magic dish.

## 30 / 甘蓝

*Brassica oleracea*
十字花科 Brassicaceae　芸薹属 *Brassica*
别名：花菜、包菜、莲花白、大头菜

### 植物形态

二年生草本。被粉霜。茎肉质。基生叶多数，质厚，扁球形，乳白色或淡绿色；基生叶及下部茎生叶长圆状倒卵形至圆形；顶端圆形，基部有宽翅叶柄，边缘有波状不明显锯齿；最上部叶长圆形，抱茎。总状花序，顶生及腋生；花淡黄色；花瓣脉纹显明；花期4月。长角果，圆柱形，果期5月。种子球形，棕色。

### 生态分布

原产中国。分布于巴基斯坦、印度、美国、日本等地。

### 生长习性

喜温和湿润、阳光充足的环境。较耐寒，适生温度15~20℃。肉质茎膨大期如遇30℃以上高温易纤维化。幼苗须在0~10℃通过春化，在长日照和适温下抽薹、开花、结果。适生于腐殖质丰富的黏壤土或砂壤土。

### 用途、价值

可食用，营养价值高。也具有很高的观赏价值，观赏期长，叶色鲜艳，用于布置花坛等，也可用于鲜切花销售。植株可药用，可预防癌细胞转移，改善血糖、血脂等。

### PLANT MORPHOLOGY

Biennial herbs, pruinose. Stems fleshy. Basal leaves numerous, thick, oblate, milky white or light green; basal and lower cauline leaves oblong-obovate to rounded; apex rounded, base with a wide-winged petiole, margin with inconspicuous sinuate serration; uppermost leaves oblong, amplexicaul. Raceme terminal, axillary; flowers light yellow; petals with obvious veins; flowering period in April. Siliqua cylindrical, fruiting period in May. Seeds spherical, brown.

### ECOLOGICAL DISTRIBUTION

Native to China. Distributed in Pakistan, India, the United States, Japan, etc.

### GROWTH HABIT

It likes a warm, humid and sunny environment. It is more cold-resistant, and the suitable growth temperature is 15-20°C. In the swelling period, the fleshy tissues of stem are prone to fibrosis in case of high temperature above 30°C. Seedlings must be vernalized at 0-10°C, it can bolt, bloom and bear fruit when in long-day condition or at the appropriate temperature. It is suitable for growing in humus-rich clay loam or sandy loam.

### APPLICATION,VALUE

It is edible and has high nutritional value. It has high ornamental value, long ornamental period and bright leaves. It is used to arrange flower beds, and can also be used for fresh cut flower sales. The plant can be used medicinally to prevent cancer cell metastasis, improve blood sugar, blood lipids, etc.

# 31 / 旱芹

*Apium graveolens*
伞形科 Apiaceae　芹属 *Apium*
别名：芹菜、胡芹

## 植物形态

二年生或多年生草本。高15~150cm。有强烈香气。根圆锥形，褐色；茎直立，并有棱角和直槽。叶片通常分裂为3小叶，小叶倒卵形，中部以上边缘疏生钝锯齿以至缺刻。复伞形花序，顶生或与叶对生，花瓣白色或黄绿色，卵圆形；花期4~7月。分生果，圆形或长椭圆形，果棱尖锐。

## 生态分布

原产地中海沿岸的沼泽地带。分布于中国、巴基斯坦、美国等地。

## PLANT MORPHOLOGY

Biennial or perennial herb. Plant height 15-150cm. Strong aroma. Root conical and brown; stems erect, with angular and straight grooves. Leaves usually divided into 3 leaflets, leaflets obovate, and edges above the middle part sparsely serrated or even notched. Compound umbels, terminal or opposite to leaf, white or yellow-green petals, ovoid, flowering period from April to July. Mericarp, round or oblong, with sharp edges.

## ECOLOGICAL DISTRIBUTION

Native to the marshes along the Mediterranean coast. Distributed in China, Pakistan, the United States, etc.

## 生长习性

耐寒性蔬菜，喜冷凉、湿润的环境，在高温干旱的条件下生长不良。耐阴，苗期需覆盖遮阳网，喜中等强度光照，生长期需充足光照。属浅根系蔬菜，吸水能力弱，耐旱力弱，蒸腾量大，对水分要求较严格。

## 用途、价值

可食用，富含多种营养元素。可药用，芹菜素具有抗氧化、降血压、扩血管等作用；钙、磷含量较高，可增强骨骼健康；纤维素含量高，可以抑制肠内细菌，预防结肠癌；富含钾，可预防浮肿。

## GROWTH HABIT

Cold-resistant vegetable, favoring cold and humid environment and does not grow well under high temperature and drought conditions. Shade-tolerant, shade nets should be covered during the seedling stage, moderate light is preferred, and sufficient light is required during the growth period. Shallow-rooted vegetable with weak water absorption ability, weak drought tolerance, large evaporation, and stricter water requirements.

## APPLICATION,VALUE

Edible and rich in many nutrients. Used as medicinally. Apigenin has the effects of anti-oxidation, lowering blood pressure and expanding blood vessels; high calcium and phosphorus content can enhance bone health; high vitamin content can inhibit intestinal bacteria and prevent colon cancer; rich in potassium, which can prevent edema.

# 32 / 茄

## 植物形态

　　直立分枝草本。高0.3~1m。小枝、叶柄及花梗均被星状茸毛，小枝多紫色，或有皮刺。叶大，卵形至长圆状卵形，边缘浅波状或深波状圆裂，叶柄长2~5cm。花单生，花萼近钟形，被长小刺，裂片披针形；花冠辐状，白色或淡紫色，裂片三角形。果形变异极大，色泽多样。

## 生态分布

　　原产亚洲热带地区。分布于亚洲、非洲、地中海沿岸、欧洲中南部、中美洲等地。

## PLANT MORPHOLOGY

　　Erect branching herb. Plant height 30-100cm. Branchlets, petioles and pedicels all covered with star-shaped hairs, and branchlets mostly purple or have prickles. Leaves large, ovate to oblong-ovate, edge shallow-wave or deep-wave circular cracks, and petiole about 2-5cm in length. Flowers solitary, calyx nearly campanulate, long thorned, lobes lanceolate; corolla radial, white or lavender, lobes triangular. Fruit shape various, a color diverse.

## ECOLOGICAL DISTRIBUTION

　　Native to tropical Asia. Distributed in Asia, Africa, the Mediterranean coast, central and southern Europe, Central America, etc.

## 生长习性

喜高温，种子发芽适温25~30℃。适生于富含有机质、保水保肥力强的土壤。对氮肥的要求较高，缺氮时延迟花芽分化，在盛花期氮不足，短柱花变多，影响植株发育。

## 用途、价值

可作蔬菜食用，富含蛋白质、维生素等多种营养成分，可防止微血管破裂出血；含多种生物碱，可抑制消化道肿瘤细胞的增殖；纤维中所含的维生素C和皂草苷，具有降低胆固醇的功效；所含B族维生素对慢性胃炎有一定辅助治疗作用。

## GROWTH HABIT

Favoring high temperature, suitable for temperature for seed germination is 25-30°C. Favored growing in organic and strong in water and fertilizer retention. Strict to the nitrogen fertilizer. Flower bud differentiation is delayed when nitrogen is deficient. In the full bloom period, nitrogen is insufficient and short-styled flower increase, which affects plant development.

## APPLICATION,VALUE

Edible as vegetables, variable nutrition such as protein and vitamin and other nutrients which can prevent capillary rupture and bleeding; contains a variety of alkaloids, which can inhibit the proliferation of gastrointestinal tumor cells; the vitamin C and saponin contained in the fiber have effect in lowering cholesterol; the contained B vitamins have a certain auxiliary therapeutic effect on chronic gastritis.

# 33 / 大豆

*Glycine max*

豆科 Leguminosae　大豆属 *Glycine*

别名：菽、黄豆塔菜

## 植物形态

一年生草本。高30~90cm。茎粗壮、直立，具棱，密被褐色长硬毛。叶常具3小叶，纸质，被黄色柔毛，小叶宽卵形或近圆形；小叶柄被黄褐色长硬毛。总状花序，蝶状花冠，紫色、淡紫色或白色；花期6~7月。荚果，长圆形，下垂，黄绿色，长4~7.5cm，宽8~15mm，密被黄褐色长毛；果期7~9月。种子椭圆状近球形，淡绿、黄、褐和黑色等，种脐明显。

## 生态分布

原产中国。分布于中国、巴基斯坦、巴西、阿根廷等地。

## 生长习性

喜温植物。适生温度20~25℃，低温下结荚延迟，低于14℃不能开花。开花期要求土壤含水量在70%~80%，否则花蕾易脱落。开花前吸肥量不到总量的15%，而开花结荚期达总吸肥量的80%以上。

## 用途、价值

可食用，加工成豆腐、豆浆、腐竹等豆制品，豆粉则是高蛋白食物，可制成多种食品，包括婴儿食品，亦是优质的蛋白饲料。还可提炼大豆异黄酮。

## PLANT MORPHOLOGY

Annual herb. Plant height 30-90cm. Stem stout, erect, angular, densely covered with brown bristles. Leaves often with 3 leaflets, papery, yellow pilose, leaflets broadly ovoid and nearly round; petioles yellowish brown with long bristles. Racemes, butterfly-shaped corolla, purple, lavender or white, flowering period from June to July. Pods, oblong, drooping, yellow-green, 40-75mm in length, 8-15mm in width, densely covered with yellowish brown hairy, fruiting period from July to September. Seeds oval and nearly spherical, light green, yellow, brown, and black, with obvious hilum.

## ECOLOGICAL DISTRIBUTION

Native to China. Distributed in China, Pakistan, Brazil, Argentina, etc.

## GROWTH HABIT

Thermophilus-favored plants. Suitable for temperature is 20-25°C, pod setting will be delayed at low temperature, and flowering will not be possible below 14°C. During the flowering period, the soil moisture content is required to be 70%-80%, otherwise the flower buds are easy to fall off. The amount of fertilizer absorbed before flowering is less than 15% of the total, and the amount of fertilizer absorbed during the flowering and podding period is more than 80% of the total.

## APPLICATION,VALUE

Edible, processed into tofu, soy milk, yuba and other soy products. Soy flour is a high-protein food that can be made into a variety of foods, including baby food, and high-quality protein feed. Extract soy isoflavones.

# 34 / 生菜

*Lactuca sativa*
菊科 Asteraceae　莴苣属 *Lactuca*
别名：玻璃菜、叶用莴苣、鹅仔菜、莴仔菜

## 植物形态

一年生或二年草本。高25~100cm。茎直立，单生。基生叶，不分裂，倒披针形、椭圆形，基部半抱茎，边缘波状或有细锯齿。头状花序，排列成圆锥花序。总苞卵球形，长1.1cm，宽6mm。瘦果，倒披针形，长4mm，宽1.3mm，压扁，浅褐色。花果期2~9月。

## 生态分布

原产欧洲地中海沿岸。分布于中国、巴基斯坦等地。

## PLANT MORPHOLOGY

Annual or biennial herb. Plant height 25-100cm. Stem erect, solitary, basal leaves, undivided, oblanceolate, elliptical, Semi-embracing stem at base, wavy or serrated at the edge. Flower heads, arranged in panicles. Involucre ovoid, 11mm in length and 6mm in width. Achenes, oblanceolate, 4mm in length and 1.3mm in width, flattened, light brown. Flowering and fruiting period from February to September.

## ECOLOGICAL DISTRIBUTION

Native to the Mediterranean coast of Europe. Distributed in China, Pakistan, etc.

## GROWTH HABIT

Favoring shady and cool environment, not

## 生长习性

　　喜冷凉环境，不耐寒，不耐旱，适生温度15~20℃；种子发芽适温18~22℃。适生于富含有机质、保水保肥力强的黏质壤土。

## 用途、价值

　　可食用。富含膳食纤维等营养物质，有消除多余脂肪的作用。亦可药用，茎叶中含有莴苣苦素，有清热提神、降低胆固醇、辅助治疗神经衰弱等功效，也可抑制癌细胞生长。

cold-resistant and not drought-resistant. Suitable temperature for growth is 15-20°C; suitable temperature for seed germination is 18-22°C. Favored growing in organic matter and strong in water and fertilizer retention with clay loam.

### APPLICATION,VALUE

Edible. Rich in dietary fiber and other nutrients, Effect in eliminating excess fat. Applicable in medicine. Stems and leaves contain lactucin, which has the effects of clearing heat and refreshing, lowering cholesterol, adjuvant treatment of neurasthenia, etc. Inhibit cancer cells.

# 35 / 蒜

## 植物形态

多年生草本。高5~60cm。鳞茎球状至扁球状，肉质、瓣状的小鳞茎紧密排列，被白色至带紫色膜质鳞茎外皮。叶宽条形至条状披针形，扁平。花葶实心，圆柱状；伞形花序密具珠芽，花常淡红色；花被片披针形至卵状披针形；花期7月。

## 生态分布

原产亚洲西部或欧洲地区。分布于中国、巴基斯坦等地。

## 生长习性

适生于土壤疏松、排水良好、有机质丰富的砂壤土。

## 用途、价值

花葶和鳞茎均供食用。含有丰富的含硫化合物，所含大蒜素具有杀菌、抑菌、抗癌等作用。鳞茎还可作药用，有温中健胃、消食理气等功效。可深加工将其转化为医药和保健品。大蒜加工废水污染也是环境保护须解决的问题。

## PLANT MORPHOLOGY

Perennial herb. Plant height 5-60cm. Bulbs globose to oblate, fleshy, petal-like small bulbs closely arranged, covered with white to purple membranous bulbs. Leaves broadly striped to strip-lanceolate and flat. Scape solid and cylindrical; umbels e densely with bulbous buds, and flowers often light red; tapels lanceolate to ovate-lanceolate; flowering period in July.

## ECOLOGICAL DISTRIBUTION

Native to Western Asia or Europe. Distributed in China, Pakistan, etc.

## GROWTH HABIT

Favoring growing in organic matter, loose, sandy loam with good drainage and rich.

## APPLICATION,VALUE

Scape and bulbs edible. Rich in sulfur-containing compounds, and allicin has the functions of bactericidal, antibacterial and anti-cancer. Its bulbs can also be used for medicinal purposes, and effect in warming the middle and invigorating the stomach, eliminating food and regulating qi. It can be further processed into medicines and health products. Garlic processing wastewater pollution is also a problem to be solved in environmental protection.

# 36 / 韭

## 植物形态

多年生草本。高25~60cm。具横生根状茎。鳞茎簇生，近圆柱状，外皮暗黄色至黄褐色，呈近网状。叶条形，扁平，实心，宽1.5~8mm。花葶圆柱状，常具2纵棱，下部被叶鞘；伞形花序半球状或近球状，具多花；花白色；花被片常具绿色或黄绿色的中脉。花果期7~9月。

## 生态分布

原产亚洲东南部。分布于中国、巴基斯坦、日本、朝鲜、韩国、越南、泰国等地。

## PLANT MORPHOLOGY

Perennial herb. Plant height 25-60cm. Transverse rhizomes. Bulbs clustered, nearly cylindrical, and outer skin dark yellow to yellowish brown, nearly reticulate. Leaves strip-shaped, flat, solid, and 1.5-8mm in width. Scape terete, often with 2 longitudinal edges, and lower part sheathed; umbels hemispherical or sub globose, with many flowers; flowers white; tepals often with green or yellow-green midribs. Flowering and fruiting period from July to September.

## ECOLOGICAL DISTRIBUTION

Native to Southeast Asia. Distributed in China, Pakistan, Japan, North Korea, South Korea, Vietnam, Thailand, etc.

## 生长习性

喜冷凉，适生温度15~25℃。适应性强，适宜空气相对湿度60%~70%。对土壤要求不严，需肥量大，耐肥力强。

## 用途、价值

韭的叶、花葶和花均作蔬菜食用；种子可入药。

## GROWTH HABIT

Favoring shade and coolness, suitable temperature for growth is 15-25°C. Strong adaptability, suitable for air relative humidity 60%-70%. Lax requirements on the soil, a large amount of fertilizer, and strong fertilizer tolerance.

## APPLICATION,VALUE

Leaves, scape and flowers edible as vegetable; seeds applicable in medicine.

# 37 / 葱

*Allium fistulosum*
石蒜科 Amaryllidaceae　葱属 *Allium*
别名：北葱

## 植物形态

多年生草本。鳞茎单生，圆柱状，稀为基部膨大的卵状圆柱形；直径1~4cm，外皮白色，稀淡红褐色。叶圆筒状，中空。伞形花序球状，多花，较疏散；花葶圆柱状，中空，中部以下膨大；花白色；花被近卵形，具反折尖头；花丝锥形；花果期4~7月。

## 生态分布

原产中国。分布于印度、韩国、日本、巴基斯坦等地。

## PLANT MORPHOLOGY

Perennial herb. Bulbs solitary, cylindrical, few ovate-cylindrical with enlarged base; 1-4cm in diameter, white skin, sparse reddish brown, cylindrical leaves, hollow. Inflorescence umbel globose, flowers many, relatively sparse; scape cylindrical, hollow, swelled below the middle; flowers white; perianth nearly ovate, with reflex points; filaments cone-shaped; flowering and fruiting period from April to July.

## ECOLOGICAL DISTRIBUTION

Native to China. Distributed in India, South Korea, Japan, Pakistan, etc.

## GROWTH HABIT

Drought resistance not flood-enduring. Suitable

## 生长习性

耐旱不耐涝。适生温度13~25℃，最高可耐45℃的高温，耐寒力强。对土壤的适应性较强，适生于土层深厚、光照适宜的肥沃土壤；喜肥，对氮元素需求量较大。

## 用途、价值

为一种重要的香料和调味品。鳞茎、叶可食用；富含蛋白质、碳水化合物等营养物质，对人体有益。可药用；所含挥发油等有效成分，具有发汗散热的作用；所含大蒜素，能抵御细菌、病毒；所含果胶，有抗癌作用。

for 13-25℃ temperature and highest 45℃, strong cold resistance. strong adaptability to soil, suitable for fertile soil with deep soil layer and suitable sunlight; like fertilizer, a large demand for nitrogen.

### APPLICATION,VALUE

An important spice and seasoning. Bulbs and leaves edible; rich in nutrients such as protein and carbohydrates, beneficial to the human body, applicable in medicine; volatile oil and other active ingredients effective in sweating and heat dissipation; the allicin effective in resist bacteria and viruses; the pectin contained in chives effective in anti-cancer.

# 38 / 豇豆

## 植物形态

一年生缠绕藤本。高5~80cm。羽状复叶具3小叶；小叶卵状菱形，有时淡紫色。总状花序腋生；花冠黄白色而微带青紫色；花期5~8月。荚果，线形，稍肉质而膨胀或坚实，具多枚种子。种子长椭圆形，黄白或暗红色。

## 生态分布

原产非洲。分布于中国、巴基斯坦、埃及、利比亚等地。

## 生长习性

耐热性强，适生温度20~25℃。不

## PLANT MORPHOLOGY

Annual winding vine. Plant height 5-80cm. Pinnate compound leaves with 3 leaflets; leaflets ovate rhomboid, sometimes lavender. Inflorescence racemes axillary; yellow-white and slightly bluish-purple; flowering period from May to August. Pods, linear, slightly fleshy and swollen or firm, with many seeds. Seeds oblong, yellowish white or dark red.

## ECOLOGICAL DISTRIBUTION

Native to Africa. Distributed in China, Pakistan, Egypt, Libya, etc.

## GROWTH HABIT

Strong heat-resistant, Suitable for 20-25°C temperature, not frost-resistant. Short-day crop. The

耐霜冻。属短日照作物。结荚期要求肥水充足。对土壤适应性强，适生于排水良好、土质疏松的土壤。

pod-setting period requires sufficient fertilizer and water. Strong adaptability to soil and suitable for well-drained and loose soil.

## 用途、价值

果可作蔬菜食用，也可药用，具有理中益气、健胃补肾、解毒等功效；干豆角含有大量的植物纤维，有润肠通便的效果。

## APPLICATION,VALUE

Fruit applicable in vegetable or medicinal, effective in invigorating stomach and kidney, detoxification etc; dried beans contain a lot of plant fiber, effective in moistening the intestines and laxatives.

# 39 / 茼蒿

## 植物形态

一年生或二年生草本。高5~70cm。叶长椭圆形或长椭圆状倒卵形，长8~10cm，二回羽状分裂。头状花序单生，花梗长15~20cm；总苞片顶端膜质扩大成附片状。舌状花瘦果有突起的狭翅肋，肋间有明显间肋。管状花瘦果有椭圆形突起的肋。花果期6~8月。

## 生态分布

原产地中海。现广泛分布于亚热带、热带、温带地区。

## 生长习性

半耐寒性蔬菜。适生温度17~22℃。对光照要求不严，是短日照蔬菜。土壤相对湿度保持在70%~80%的环境下。

## 用途、价值

可作蔬菜食用，具有调胃健脾、降压补脑等效用。提取制作的茼蒿精油对害虫具有拒食性，对咳嗽痰多、脾胃不和等均有裨益，还可用于制作鱼类诱食剂。

## PLANT MORPHOLOGY

Annual or biennial herb. Plant height 5-70cm. Leaves oblong or oblong obovate, 8-10cm in length, and two pinnate splited. Inflorescence capitate, solitary, peduncle 15-20cm in length, membrane at the top of the bracts expanded into attached plates. Achene ligulate with protuberant narrow winged ribs, intercostal obvious intercostal. Tuberous achenes elliptical ribs. Flowering and fruiting period from June to August.

## ECOLOGICAL DISTRIBUTION

Native to the Mediterranean region. Now widely distributed in subtropical, tropical and temperate regions.

## GROWTH HABIT

Semi-hardy vegetable. Suitable for 17-22°C temperature. not strict to the light, short-day vegetable. The relative humidity of the soil is kept at 70%-80%.

## APPLICATION,VALUE

Applicable in vegetable edible and effective in regulating the stomach and strengthening the spleen, reducing blood pressure and tonifying the brain. The extracted chrysanthemum chrysanthemum essential oil effective in antifeedant properties to pests, beneficial to coughing and sputum, spleen and stomach discordance, etc., and also used to make fish attractants.

# 40 / 黄瓜

*Cucumis sativus*

葫芦科 Cucurbitaceae　黄瓜属 *Cucumis*

别名：青瓜、胡瓜、旱黄瓜

## 植物形态

　　一年生蔓生或攀缘草本。茎枝伸长，有棱沟，被白色糙硬毛。卷须细，不分歧，具白色柔毛。叶片宽卵形，膜质，长宽均7~20cm。叶柄被糙硬毛，长10~20cm。花常叶腋簇生；花梗纤细，被微柔毛；花萼筒狭钟状或近圆筒状，密被白色长柔毛；花冠黄白色，宿存。果实长圆形或圆柱形，长10~50cm，熟时黄绿色，表面粗糙，有具刺尖的瘤状突起，花果期夏季。种子小，狭卵形，白色，长5~10mm。

## PLANT MORPHOLOGY

Annual trailing or climbing herb. Stems and branches elongation, furrowed, covered with white rough hairs. Tendrils thin, not divided, with white pilose. Leaves broadly ovoid, membranous, 7-20cm in length and width. Petiole coarsely hairy, 10-20cm in length. Flower often leaf axillary clusters; pedicel slender, puberulent; calyx tube narrowly campanulate or nearly cylindrical, densely covered with white pilose; corolla yellow-white, persistent. Fruit oblong or cylindrical, 10-50cm in length, yellowish-green when ripe, with rough surface, with thorn-pointed tumor-like protrusions, flowering and fruiting in summer. Seeds small, narrowly ovoid, white, about 5-10mm in length.

## 生态分布

原产印度。现分布于各温带和热带地区。

## 生长习性

喜温暖，不耐寒。适生温度10~32℃。对水分需求量大。喜湿润而不耐涝，适生于富含有机质的肥沃土壤。

## 用途、价值

果实食用价值高，营养丰富，富含蛋白质、糖类、维生素等营养成分。具有除热、利尿、清热解毒的功效。

## ECOLOGICAL DISTRIBUTION

Native to India. Now distributed in various temperate and tropical regions.

## GROWTH HABIT

Thermophilic, unhardy. Suitable for 10-32℃ temperature. Large demand to water. Wet-loving but not flood-enduring. Suitable for fertile soil rich in organic matter.

## APPLICATION,VALUE

Fruit high edible value, rich nutrition, and rich in nutrients such as protein, sugar and vitamins. Effective in removing heat, diuresis, clearing away heat and detoxifying.

# 41 / 番薯

*Ipomoea batatas*
旋花科 Covolvulaceae  虎掌藤属 *Ipomoea*
别名：白薯、红苕、红薯、甜薯、地瓜

## 植物形态

一年生草本。具圆形、椭圆形或纺锤形块根。茎多分枝，圆柱形或具棱，绿或紫色，茎节易生不定根。叶片宽卵形，长4~13cm，宽3~13cm；基部心形或近于平截，叶色有浓绿、黄绿、紫绿等。聚伞花序；花冠粉红、白、淡紫或紫色，钟状或漏斗状。蒴果，卵形或扁圆形。种子通常2枚。

## 生态分布

原产南美洲及大安的列斯群岛、小安的列斯群岛。分布于中国、巴基斯坦等地。

## 生长习性

喜温植物。不耐寒，适生温度22~30℃。喜光，短日照作物。耐旱，适应性强。耐盐碱；根系发达、吸肥力强，适生于土层深厚、疏松的土壤。

## 用途、价值

块根可食用。是一种营养齐全而丰富的天然滋补食品。亦可药用，具有补虚乏、健脾胃等功效。根、茎、叶是优良的饲料。淀粉是生产增塑剂、高级吸收性树脂的重要原料。

## PLANT MORPHOLOGY

Annual herb. With round, oval or spindle-shaped roots. Stems multi-branched, cylindrical or edged, green or purple, and stem nodes suitable for adventitious roots. Leaves broadly ovate, 4-13cm in length and 3-13cm in width; base heart-shaped or nearly truncated, and leaf dark green, yellow-green, and purple-green. Inflorescence cymes; corolla pink, white, lavender or purple, bell-shaped or funnel-shaped. Capsules, ovoid or oblate. Seeds usually 2 pieces.

## ECOLOGICAL DISTRIBUTION

Native to South America and the Great Antilles and Lesser Antilles. Distributed in China, Pakistan, etc.

## GROWTH HABIT

Thermophilic, unhardy, Suitable for 22-30℃ temperature. Light-favored short-day crop. Drought resistance, strong adaptability. Salt-alkali resistance; developed root system, strong fertility absorption, suitable for deep soil and loose soil.

## APPLICATION,VALUE

Root tuber edible. Natural nourishing food with complete nutrition, applicable in medicine, effective in invigorating deficiency and strengthening the spleen and stomach. Roots, stems, and leaves are excellent feed. Starch is an important raw material for the production of plasticizers and advanced absorbent resins.

# 42 / 辣椒

*Capsicum annuum*
茄科 Solanaceae　辣椒属 *Capsicum*
别名：牛角椒、长辣椒、菜椒、灯笼椒

## 植物形态

一年生或多年生草本。高40~80cm。茎微生柔毛，分枝稍呈"字"形曲折。叶互生，枝顶端节呈双生或簇生状，矩圆状卵形、卵形或卵状披针形，长4~13cm，宽1.5~4cm；叶柄长4~7cm。花单生，俯垂；花萼杯状；花冠白色，裂片卵形。果梗较粗壮，俯垂；未成熟时绿色，成熟后呈红色、橙色或紫红色；花果期5~11月。种子扁肾形，长3~5mm，淡黄色。

## 生态分布

原产墨西哥、哥伦比亚。分布于中

## PLANT MORPHOLOGY

Annual or perennial herb. Plant height 40-80cm. Stem slightly pilose, and branches slightly zigzag-shaped. Leaves alternate, branch tips twin or clustered, oblong-ovate, ovate or ovate-lanceolate, 4-13cm in length and 1.5-4cm in width; petioles 4-7cm in length. Flowers solitary, drooping; calyx cup-shaped; corolla white, lobes ovate. Fruit stalk thick and drooping; green when immature, and red, orange or purplish red after maturity. Flowering and fruiting period from May to November. Seeds kidney-shaped, 3-5mm in length, light yellow.

## ECOLOGICAL DISTRIBUTION

Native to Mexico and Colombia. Distributed in China, Pakistan, India, etc.

国、巴基斯坦、印度等地。

## 生长习性

不耐旱也不耐涝。适生温度 15~34℃。苗期要求温度较高，不耐寒。对水分要求严格，喜干爽的空气条件。

## 用途、价值

为重要的蔬菜和调味品。果实可食用，维生素C含量高，居蔬菜之首位。亦可药用，能缓解胸腹冷痛，制止痢疾，杀抑胃腹内寄生虫，控制心脏病及冠状动脉硬化；还能刺激口腔黏膜，引起胃的蠕动，促进唾液分泌，增强食欲，促进消化等功效。种子油可食用、果有驱虫和发汗的药效。

## GROWTH HABIT

Not drought resistance and flood-enduring. Suitable for 15-34°C temperature. seedling stage high temperature quality, unhardy. Strict to the water quality, prefers dry air conditions.

## APPLICATION,VALUE

Important vegetable and condiment. Fruit edible, high vitamin C content, ranked first in vegetables. Applicable in medicine to relieve chest and abdominal pain, stop dysentery, kill stomach and abdomen parasites, control heart disease and coronary artery sclerosis; stimulate oral mucosa, cause gastric peristalsis, promote saliva secretion, enhance appetite, and promote digestion. Seed oil edible, fruit effective in expelling insects and sweating.

# 43 / 胡萝卜

*Daucus carota*
伞形科 Apiaceae 胡萝卜属 *Daucus*
别名：赛人参、红萝卜

## 植物形态

一年生或二年生草本。高 60~90cm。根粗壮，长圆锥形，呈橙红色或黄色，茎直立，多分枝。二至三回羽状复叶，有小尖头。复伞花序；花梗被糙硬毛；总苞片多数，呈叶状羽状分裂；花白色，或淡红色；花期4月。果实圆形，棱上有白刺。

## 生态分布

原产亚洲西部。分布于亚洲、欧洲和美洲地区。

## PLANT MORPHOLOGY

Annual or biennial herb. Plant height 60-90cm. Root stout, oblong cone-shaped, orange red or yellow, stem erect, multiple branches. 2-3 pinnate compound leaves, with small pointed head. Inflorescence umbrella; pedicel scabrous; bracts numerous, leaf-like pinnately divided; flowers white or reddish; flowering period in April. Fruit round with white thorns on the edge.

## ECOLOGICAL DISTRIBUTION

Native to western Asia. Distributed in Asia, Europe and America.

## GROWTH HABIT

The suitable temperature for growth is 20-25°C. Light-favored. It is suitable for loam or sandy loam

## 生长习性

适生温度20~25℃。喜光。适生于土层深厚肥沃、排水良好的壤土或砂壤土。

## 用途、价值

根可食用。营养价值高，富含维生素，可刺激皮肤新陈代谢，增加血液循环。胡萝卜素在人体内能转化成维生素A，有防癌的作用；叶可防治水痘与急性黄疸型肝炎；胡萝卜汁可预防夜盲症、眼干燥症等。

with deep and fertile soil layer and good drainage.

### APPLICATION,VALUE

The root is edible. High nutritional value, rich in vitamins, can stimulate skin metabolism, increase blood circulation. Carotene can be converted into vitamin A in human body, which has the effect of preventing cancer; leaves can prevent varicella and acute jaundice hepatitis; carrot juice can prevent night blindness and dry eye disease.

# 44 / 薤菜

*Ipomoea aquatica*
旋花科 Convolvulaceae  虎掌藤属 *Ipomoea*
别名：通菜蓊、空心菜、通心菜

## 植物形态

一年生草本。蔓生或漂浮于水。茎有节，节间中空，节上生根。叶卵形或长卵状披针形，长3.5~17cm，宽0.9~8.5cm，具小短尖头，基部心形、戟形或箭形，有时基部有少数粗齿，两面近无毛或偶有稀疏柔毛；叶柄长3~14cm。聚伞花序腋生，基部被柔毛；花冠白色、淡红色或紫红色，漏斗状。蒴果，卵球形至球形。种子密被短柔毛。

## 生态分布

原产中国。分布于亚洲、非洲和大洋洲等地。

## PLANT MORPHOLOGY

Annual herb. Creeping or floating in water. Stem with nodes, internodes hollow with roots on. Leaves are ovate or long ovate lanceolate, 3.5-17cm in length and 0.9-8.5cm in width, with small spikes, base heart-shaped, halberd shaped or arrow shaped, sometimes with a few coarse teeth at the base, both sides subglabrous or occasionally sparsely pilose; petiole 3-14cm in length. Cymes, axillary, base pilose; corolla white, reddish or purplish red, funnel-shaped. Capsule ovoid to globose. Seeds or densely pubescent.

## ECOLOGICAL DISTRIBUTION

Native to China. Distributed in Asia, Africa and Oceania, etc.

## 生长习性

不耐寒；遇霜冻，茎、叶枯死。适生于气候温暖湿润、土壤肥沃多湿的环境。

## 用途、价值

可作蔬菜食用。亦可药用，内服可解饮食中毒，外敷治骨折、腹水及无名肿毒。也是一种优良饲料。

## GROWTH HABIT

Unhardy; in case of frost, the stems and leaves die off. It is suitable for warm and humid climate, fertile and humid soil.

## APPLICATION,VALUE

It can be used as vegetable. It can also be used for medicine. It can be taken orally to solve food poisoning, and external application for fracture, ascites and unknown swelling. It is also a good feed.

# 45 / 银灰旋花

*Convolvulus ammannii*
旋花科 Convolvulaceae　旋花属 *Convolvulus*
别名：小旋花、亚氏旋花、彩木

## 植物形态

多年生草本。高2~10cm。根状茎短，密被柔毛。叶互生，呈线形或狭披针形，长1~2cm，宽1~4mm，被银白色茸毛。花单生枝端，具细花梗；花冠小，漏斗状，淡玫瑰色或白色带紫色条纹，有毛；花期6~8月。蒴果，球形，长4~5mm；果期7~9月。种子2~3枚，卵圆形，具喙，淡褐红色。

## 生态分布

原产中国。分布于巴基斯坦、印度等地。

## PLANT MORPHOLOGY

Perennial herb. Plant height 2-10cm. Rhizome short, densely pilose. Leaves alternate, linear or narrowly lanceolate, 1-2cm in length and 1-4mm in width, covered with silvery white tomentose. Flowers solitary at the end of branches, with fine pedicels; corolla small, funnel-shaped, light rose or white, with purple stripes, hairy; flowering period from June to August. Capsule, globose, about 4-5mm in length; fruiting period from July to September. Seeds 2-3, ovoid, beaked, reddish brown.

## ECOLOGICAL DISTRIBUTION

Native to China. Distributed in Pakistan, India, etc.

## 生长习性

较耐高温，喜光，喜干旱排水好的环境。耐旱，耐盐碱，对土壤要求不严，能在沙漠中生长。

## 用途、价值

是盐碱化草地和各类草原群落常见的伴生种，是草地退化和盐碱化的优势种。还可作饲料。全草可入药，主治感冒、咳嗽等。

## GROWTH HABIT

Heat-resistant, light-favored. It likes dry and well drained growth environment. Drought and saline-alkali resistance, has no strict requirements on soil, and can grow in desert.

## APPLICATION,VALUE

It is the common associated species of saline alkali grassland and various grassland communities, and is the dominant species of grassland degradation and salinization. Used as feed. The whole herb can be used as medicine, mainly for cold and cough.

# 46 / 时钟花

西番莲科 Passifloraceae

*Turnera ulmifolia*

时钟花属 *Turnera*

## 植物形态

多年生草本。高30~80cm。叶互生，椭圆形至倒阔披针形，边缘有锯齿，叶基有一对腺体。花腋生，花冠金黄色，5瓣，每朵花至午前凋谢；花期春夏季。蒴果，室背开裂或不开裂。种子多数。

## 生态分布

原产美洲热带。分布于中国、印度、巴基斯坦等热带地区。

## 生长习性

喜高温。适生温度20~30℃。喜

## PLANT MORPHOLOGY

Perennial herb. Plant height 30-80cm. Leaves alternate, elliptic to oblong lanceolate, margin serrate, leaf base with a pair of glands. Flowers axillary, corolla golden yellow, petals 5, each flower to wither before noon; flowering in spring and summer. Capsule, loculicidally dehiscent or not dehiscent. Seeds numerous.

## ECOLOGICAL DISTRIBUTION

Native to tropical America. Distributed in China, India, Pakistan and other tropical areas.

## GROWTH HABIT

It likes high temperature. The suitable temperature for growth is 20-30℃. light-favored and slightly drought tolerant. It is suitable for fertile

光，略耐旱。适生于肥沃壤土和富含腐
殖质的土壤。

## 用途、价值

观赏价值高，在园林中应用广泛。
时钟花不休息、不迟到，在古时有一定
的计时作用。全草可入药。有轻微毒
性，具有清热解毒、驱虫等功效。

loam and humus soil.

## APPLICATION,VALUE

It has high ornamental value and is widely used in gardens. The clock flower does not rest or arrive late, and has a certain timing function in ancient times. The whole herb can be used as medicine. It has slight toxicity, and has the effects of clearing away heat and detoxification, expelling insects and so on.

# 47 / 鸡冠花

## 植物形态

一年生直立草本。高30~80cm。茎绿色或红色，有棱纹凸起。叶互生，具柄；叶片长5~13cm，全缘。穗状花序，圆锥状矩圆形，表面羽毛状；花被片红色、紫色、黄色、橙色或红黄色相间；花果期7~9月。胞果，卵形，熟时盖裂，包于宿存花被内。种子肾形。

## 生态分布

原产非洲、美洲热带。分布于中国、巴基斯坦、巴西等地。

## PLANT MORPHOLOGY

Annual erect herb. Plant height 30-80cm, green or red, with ridges. Leaves alternate, stipitate; leaf blade 5-13cm in length, entire. Spike inflorescence, conical, oblong, feathery on the surface; perianth red, purple, yellow, orange or red yellow; flowering and fruiting period from July to September. Utricle, ovate, covered with persistent perianth when ripe. Seeds reniform.

## ECOLOGICAL DISTRIBUTION

Native to Africa and tropical America. Distributed in China, Pakistan, Brazil, etc.

## GROWTH HABIT

Favoring warm and dry climate, not drought resistant, Light-favored, not waterlogging. The soil

## 生长习性

　　喜温暖干燥气候，不耐旱，喜光，不耐涝。对土壤要求不严，一般土壤均可种植。

## 用途、价值

　　对二氧化硫、氯化氢有良好的抗性，可起到绿化、美化和净化环境的多重作用，是一种抗污染环境的观赏花卉。花和种子可药用，作收敛剂，有止血、止泻等功效。

requirement is not strict, the general soil can be planted.

## APPLICATION,VALUE

It has good resistance to sulfur dioxide and hydrogen chloride, and can play a multiple role in greening, beautifying and purifying the environment. It is a kind of ornamental flower with anti pollution environment. Flowers and seeds can be used as medicine, as astringent, have hemostasis, antidiarrheal and other effects.

# 48 / 仙人掌

*Opuntia dillenii*

仙人掌科 Cactaceae　仙人掌属 *Opuntia*

别名：仙巴掌、霸王树、火焰、火掌、牛舌头

## 植物形态

多年生草本。高1~3m。上部分枝宽倒卵形、倒卵状椭圆形或近圆形，边缘呈不规则波状，绿色或蓝绿色，无毛；成长后刺增粗并增多，密生短绵毛和倒刺刚毛；刺粗钻形，坚硬；倒刺刚毛暗褐色，直立；叶钻形，早落。花辐状，直径5~7cm；疏生突出的小窠，小窠具短绵毛、倒刺刚毛和钻形刺；瓣状花被片倒卵形或匙状倒卵形，全缘或浅啮蚀状；花期6~12月。浆果倒卵球形，长4~6cm，宽2.5~4cm，紫红色。种子扁圆形，长4~6mm，宽4~5mm，淡黄褐色。

## 生态分布

原产墨西哥东海岸、美国南部等地。分布于中国、印度、巴基斯坦和澳大利亚等地。

## 生长习性

喜温暖，喜光照充足的环境，耐旱，不耐寒，忌涝。适生于中性、微碱性土壤。

## 用途、价值

作围篱，有很高的观赏价值，在园林中用途广泛。浆果酸甜多汁，可鲜食，也可加工罐头、饮料；茎供药用，内服外用治疗多种疾病。可作饲料，在干旱地区成片种植，作为牲畜的饲料源。

## PLANT MORPHOLOGY

Perennial herb. Plant height 1-3m. Upper branches wide, obovate, oval or nearly round, with irregular wavy edges, green or blue-green, glabrous; After growing, thorns thickened and increased, densely producing short woolly hairs and barb bristles; spines coarsely subulate, hard; barbed setae dark brown, erect. Leaves subulate, caducous. Flowers radiate, 5-7cm in diameter; sparsely protruding small nests with short woolly hairs, barbed setae and subulate spines; petal perianth obovate or spatulate obovate, entire or shallow erose; flowering period from June to December. Berry, obovoid, 4-6cm in length, 2.5-4cm in width, purplish red. Seeds oblate, 4-6mm in length, 4-5mm in width, yellowish brown.

## ECOLOGICAL DISTRIBUTION

Native to the east coast of Mexico and the southern United States. Distributed in China, India, Pakistan, Australia, etc.

## GROWTH HABIT

Favoring warm, well-lit environment. It is resistant to drought, cold and waterlogging. It is suitable for neutral and slightly alkaline soil.

## APPLICATION,VALUE

As a fence, it has high ornamental value and is widely used in gardens. The berries are sour, sweet and juicy, and can be eaten fresh. They can also be used to process canned and beverage. The stems are used for medicine, and for internal and external use to treat various diseases. It can be used as feed, and can be planted in dry area as feed source for livestock.

# 49 / 白接骨

*Asystasia neesiana*

爵床科 Acanthaceae　十万错属 *Asystasia*

别名：玉龙盘

## 植物形态

草本。高0.2~1m。具白色，富黏液；竹节形根状茎，呈四棱形。叶卵形至椭圆状矩圆形，长5~20cm，边缘微波状至具浅齿，侧脉两面凸起，疏被微毛。总状花序，顶生，长6~12cm；花单生；主花轴和花萼被腺毛；花冠淡紫红色，漏斗状，花冠筒细长，长3~4cm。蒴果，长18~22mm；上部具4枚种子，下部实心细长似柄。

## 生态分布

分布于中国、巴基斯坦、印度、越南、缅甸等地。

## PLANT MORPHOLOGY

Herb. Plant height 0.2-1m. White, rich in mucus; bamboo-shaped rhizomes, quadrangular in shape. leaves ovate to elliptic and oblong, 5-20cm in length, with microwave-like edges to shallow teeth, lateral veins convex on both sides, and sparse puberulent. Racemes, terminal, 6-12cm in length; flowers solitary; main floral axis and calyx glandular hairs; corolla pale purple-red, funnel-shaped, corolla tube slender, 3-4cm in length. Capsule, 18-22mm in length; upper part with 4 seeds, lower part solid and stalk-like slender.

## ECOLOGICAL DISTRIBUTION

Distributed in China, Pakistan, India, Vietnam, Myanmar, etc.

## 生长习性

生于山坡、山谷林下阴湿的石缝内和草丛中，溪边亦有。

## 用途、价值

观赏价值较高，可引入园林作为宿根花卉，对阴湿环境适应性好。叶和根状茎可入药，具活血化瘀的功效，可治疗跌打损伤。

## GROWTH HABIT

It grows in the wet stone crevices and grass under the hillside and valley forest, and also by the stream.

## APPLICATION,VALUE

It has high ornamental value and can be introduced into gardens as perennial flowers, which has good adaptability to humid environment. Leaves and rhizomes can be used as medicine, which has the effects of promoting blood circulation, removing blood stasis and treating traumatic injury.

# 50 / 天人菊

菊科 Asteraceae 天人菊属 *Gaillardia*

*Gaillardia pulchella*

别名：虎皮菊

## 植物形态

一年生草本。高20~60cm。茎多分枝，分枝斜升，被短柔毛或锈色毛。叶匙形或倒披针形，长5~10cm，基部无柄或心形半抱茎，叶两面被伏毛。头状花序径5cm；背面有腺点，基部密被长柔毛；舌状花黄色，基部带紫色，长1cm；管状花裂片三角形，被节毛。瘦果，长2mm；花果期6~8月。

## 生态分布

原产于热带美洲。分布于中国、巴基斯坦等地。

## PLANT MORPHOLOGY

Annual herb. Plant height 20-60cm. Stems more branched, branches oblique rise-up, puberulent or rust-colored hairs. Leaves spoon-shaped or oblanceolate, 5-10cm in length, sessile or heart-shaped at the base, leaves covered with hairs on both sides. head-like flower 5cm in diameter; gland spots on the back, and base densely pilose; tongue-like flower yellow, base purple, 1cm in length; tubular flower lobes triangular, with nodules. Achene, 2mm in length; flowering and fruit period from June to August.

## ECOLOGICAL DISTRIBUTION

Native to tropical America. Distributed in China, Pakistan, etc.

## 生长习性

喜温植物。耐热，不耐寒，喜阳光充足，略耐半阴。喜湿，对土壤的肥水要求较高。

## 用途、价值

观赏价值高，色彩艳丽，花期长，易栽培，可用作花坛、花丛的材料，园林应用广泛；生命力强，全株有柔毛，可防止水分散失，是良好的防风固沙植物。

## GROWTH HABIT

Thermophilic plant. Heat resistant, unhardy, light-favored, slightly resistant to half shade. It is wet and requires more fertilizer and water for the soil.

## APPLICATION, VALUE

High ornamental value, bright colors, long flowering period, easy cultivation, can be used as materials for flower beds and flowers, and is widely used in gardens; strong vitality, the whole plant is pilose, which can prevent water loss and is a good wind-proof and sand-fixing plant.

# 51 / 黄金菊

*Chrysanthemum morifolium*
菊科 Asteraceae  菊属 *Chrysanthemum*
别名：小白菊、小汤黄、杭白菊

## 植物形态

多年生草本。高0.6~1.5m。茎直立，被柔毛。叶互生，羽状浅裂或半裂，下面被白色短柔毛，边缘有粗大锯齿或深裂。头状花序；总苞片多层，被柔毛；舌状花白色、红色、紫色或黄色；花期9~11月，果实多不发育。

## 生态分布

原产中国。现广泛分布于亚热带、热带、温带地区。

## 生长习性

喜光，较耐寒。喜温暖湿润气候。

## PLANT MORPHOLOGY

Perennial herb. Plant height 0.6-1.5m. Stem erect, pilose. Leaves alternate, pinnately lobed or lobed, white puberulent below, margin coarsely serrate or deeply lobed. Flower white, red, purple or yellow, flowering period from September to November, fruit mostly undeveloped.

## ECOLOGICAL DISTRIBUTION

Native to China. Wide distributed in subtropical, tropical and temperate regions.

## GROWTH HABIT

Light-favored, cold-resistant, favoring warm and humid climate. Likes warm and humid climate. Suitable for growth is 20-25°C temperature. Short-day plants; suitable for sandy loam with deep soil,

适生温度20~25℃。短日照植物；适生于土层深厚、富含腐殖质、排水良好的砂壤土。

## 用途、价值

生长旺盛，萌发力强，观赏价值高。花可食用。亦可入药，久服或饮菊花茶能调节血液循环等。

rich in humus, and good drainage.

### APPLICATION,VALUE

Strong growth, strong germination and high ornamental value. Flowers are edible. It can also be used as medicine. Taking or drinking chrysanthemum tea for a long time can regulate blood circulation.

# 52 / 百日菊

*Zinnia elegans*
菊科 Asteraceae　百日菊属 *Zinnia*
别名：火球花、对叶菊、秋罗、步登高

## 植物形态

一年生草本。高0.5~1m。茎直立，被糙毛或硬毛。叶宽卵圆形或长圆状椭圆形，基部稍心形抱茎。头状花序，总苞宽钟状，边缘黑色；托片附片紫红色，流苏状三角形；舌状花深红、玫瑰、紫堇或白色，先端齿裂或全缘；管状花黄或橙色，密被黄褐色茸毛；花期6~9月。瘦果被密毛，倒卵状楔形，果期7~10月。

## 生态分布

原产墨西哥。分布于中国、巴基斯坦、印度等地。

## PLANT MORPHOLOGY

Annual herb. Plant height 0.5-1m. Stem erect, coarsely or bristly hairy. Leaves broadly ovoid or oblong-elliptic, and base slightly heart-shaped. Inflorescence head, involucre broadly bell-shaped, with black margins; pallets attached with purplish red, fringe-shaped triangles; tongue-shaped flowers crimson, rose, pansy or white, apex teeth cracked or entire; tubular flowers yellow or orange, densely covered with yellowish brown hairy; flowering period from June to September. Achenes pilose, obovate and wedge-shaped, fruiting from July to October.

## ECOLOGICAL DISTRIBUTION

Native to Mexico. Distributed in China, Pakistan, India, etc.

## 生长习性

喜温、不耐寒；适生温度15~30℃，喜光、耐干旱、耐瘠薄、忌连作；根深茎硬，不易倒伏，适生于肥沃深厚土壤。

## 用途、价值

观赏价值高，花大色艳，株型美观，可用于花坛、花境、花带。亦可药用，有清热、解毒等功效。

## GROWTH HABIT

Thermophilic and unhardy; suitable for 15-30°C temperature, light-favored, drought-resistant, barren, and avoids continuous cropping; it has deep roots and hard stems, not easy to lodging, and is suitable for growing in fertile deep soil.

## APPLICATION,VALUE

High ornamental value, large and colorful flowers, beautiful plant shape, can be used in flower beds, flower borders and flower belts. It can also be used as medicine, and has the effects of clearing away heat and detoxifying.

# 53 / 万寿菊

*Tagetes erecta*

菊科 Asteraceae　万寿菊属 *Tagetes*

别名：臭芙蓉

## 植物形态

一年生草本。高0.5~1.5m。茎直立，粗壮，具纵细条棱，分枝向上平展。叶羽状分裂，长5~10cm，宽4~8cm，裂片长椭圆形或披针形，具锐齿。头状花序，单生，径5~8cm，花序梗顶端棍棒状；总苞长1~2cm，杯状；舌状花黄或暗橙黄色，长3cm，舌片倒卵形，长1.4cm，基部呈长爪状；管状花花冠黄色，长约10mm。瘦果，线形，被微毛。

## 生态分布

原产墨西哥。分布于中国、巴基斯

## PLANT MORPHOLOGY

Annual herb. Plant height 0.5-1.5m. Stem erect, strong, with longitudinal thin ribs, branches spreading upward. Leaves pinnately divided, 5-10cm in length and 4-8cm in width. Lobes oblong or lanceolate, with sharp teeth. Head-shaped flower head, solitary, 5-8cm in diameter, peduncle top stick-shaped; involucre 1-2cm in length, cup-shaped; tongue-shaped flower yellow or dark orange yellow, 3cm in length, tongue obovate, 1.4cm in length, claws grow at the base; the tubular flower corolla yellow, about 10mm in length. Achenes, linear, puberulent.

## ECOLOGICAL DISTRIBUTION

Native to Mexico. Distributed in China, Pakistan, etc.

坦等地。

## 生长习性

喜光，适生温度15~25℃。对土壤要求不严，适生于肥沃、排水良好的砂质壤土。

## 用途、价值

园林绿化花卉，用于点缀花坛、广场等，可作盆栽或切花。对氟化氢、二氧化硫等气体有较强的抗性和吸收作用。可食用；是花卉食谱中的名菜。花有香味，可作芳香剂。亦可药用，有解毒消肿等功效。

## GROWTH HABIT

Light-favored and the suitable for growth is 15-25°C temperature. The soil is not strict, suitable for fertile, well drained sandy loam.

## APPLICATION,VALUE

It is a landscaping flower, used to decorate flower beds, squares, etc., and can be used as potted plants or cut flowers. It has strong resistance and absorption to hydrogen fluoride, sulfur dioxide and other gases. Edible; it is a famous dish in flower recipes. Flowers are fragrant and can be used as fragrances. It can also be used as medicine, and has the effects of detoxification and detumescence.

# 54 / 羽芒菊

*Tridax procumbens*
菊科 Asteraceae　羽芒菊属 *Tridax*

## 植物形态

多年生铺地草本。节处常生多数不定根，长30~100cm，基部略呈四方形，被倒向糙毛或脱毛。基部叶略小，花期凋萎；中部叶有柄，叶片披针形或卵状披针形，长4~8cm，宽2~3cm，基生三出脉；基部近楔形，边缘有粗齿或基部近浅裂。头状花序；花序梗被白色疏毛；花托稍突起；花多数，花冠管状，被短柔毛，边缘有时带波浪状；花期11月至翌年3月。瘦果，陀螺形或倒圆锥形，被疏毛。

## PLANT MORPHOLOGY

Perennial ground covering herbs. Most adventitious roots often found at the nodes, 30-100cm in length, slightly square at the base, downwardly hispid or depilated. Basal leaves slightly small and withered at flowering stage; middle leaves petiole, the leaves lanceolate or ovate lanceolate, 4-8cm in length and 2-3cm in width, with three veins at the base; base nearly cuneate, with coarse teeth at the edge or nearly lobed at the base. Inflorescence head; peduncle white; receptacle slightly protuberant; flowers numerous, corolla tubular, pubescent, margin sometimes undulate; flowering period from November to March of the following year. Achenes, gyrate or obconical, with hirsute.

## 生态分布

原产中国。分布于巴基斯坦、印度、泰国、越南及巴西等地。

## 生长习性

喜光植物，极耐干旱，在耕作地、空旷的荒地上生长良好；在干燥或湿润的沙地生长最盛。对土壤要求不严。

## 用途、价值

富含碳水化合物、粗蛋白质和粗纤维，可作饲料，草质柔嫩多汁，在冬春干旱季节，多数牧草枯黄老化时，仍青绿柔嫩。易蔓延成片，危害农作物，降低植物多样性。

## ECOLOGICAL DISTRIBUTION

Native to China. Distributed in Pakistan, India, Thailand, Vietnam, Brazil, etc.

## GROWTH HABIT

The positive plant, extremely resistant to drought, grows well on cultivated land and open wasteland, and grows most in dry or humid sandy land. Not strict to soil quality.

## APPLICATION, VALUE

Rich in carbohydrates, crude protein and crude fiber, it can be used as feed, and the grass is tender and juicy. In the dry season of winter and spring, most grasses are still green and tender when they are yellow and aging. It is easy to spread into pieces, harm crops and reduce plant diversity.

# 55 / 南美蟛蜞菊

*Sphagneticola trilobata*
菊科 Asteraceae　蟛蜞菊属 *Sphagneticola*
别名：三裂叶膨蜞菊、地锦花、穿地龙

## 植物形态

多年生草本。矮小，匍匐状，被短而压紧的毛。叶对生，叶上有3裂，叶缘有锯齿，叶片被刚毛，紫红色或绿色；主脉3条。头状花序，具长柄，花序直径约2cm；花托扁平；边缘舌状花，黄色；中央管状花。瘦果，有棱，先端有硬冠毛；果期夏秋季。

## 生态分布

原产热带美洲中南部。在中国、泰国、越南、巴基斯坦等地有引种栽培。

## PLANT MORPHOLOGY

Perennial herb. Short, creeping, with short compressed hairs. Leaf opposite, three lobed, leaf margin with the serration, leaf blade covered with setae, purplish red or green; main vein 3. Inflorescence head, long petiole, inflorescence about 2cm in diameter; receptacle flat; marginal ligulate, yellow; central tubular flower. Achenes, angulate, apex bristly; fruiting period in summer and autumn.

## ECOLOGICAL DISTRIBUTION

Native to central and southern tropical America. Introduced and cultivated in China, Thailand, Vietnam, Pakistan, etc.

## GROWTH HABIT

Light-favored, heat-resistant and drought

## 生长习性

　　喜光、喜高温、耐旱。适生温度18~30℃。对土壤要求不严。花期极长，终年可见花。

## 用途、价值

　　园林绿化植物，常作地被应用；对有害金属元素的吸附积累作用较强，是较好的污水处理、湿地恢复优选植物。可用于深加工制药、制饲料等；或分析提取其成分进行人工合成，为生物防控所用；可净化修复被城市垃圾渗滤液污染的土壤，宜作为垃圾填埋场植被重建材料。

resistance. Suitable for temperature is 18-30°C. Not strict to the soil quality. Flowering period is extremely long, and flowers can be seen all year round.

### APPLICATION,VALUE

　　Landscaping plants, used for ground greening; it has strong adsorption and accumulation of harmful metal elements, and is a better plant for sewage treatment and wetland restoration. Used for deep processing of pharmaceuticals, feed, etc.; or its components can be analyzed and extracted for artificial synthesis, used for biological control; it can purify and repair the soil polluted by urban landfill leachate, and should be used as the vegetation reconstruction material of landfill site.

# 56 / 五彩苏

*Coleus scutellarioides*
唇形科 Lamiaceae　鞘蕊花属 *Coleus*
别名：锦紫苏、洋紫苏、五色草、老来少、彩叶草

## 植物形态

直立或上升草本。茎紫色，四棱形，被微柔毛，具分枝。叶卵形，长4~13cm，圆齿状，黄、深红、紫及绿色，两面被微柔毛，下面疏被红褐色腺点。轮伞花序组成圆锥花序；花萼钟形，长2~3mm；花冠紫或蓝色，长2cm，被柔毛；花期7月。小坚果，宽卵圆形或圆形，压扁，褐色，具光泽，长1~2mm。

## 生态分布

原产中国。分布于中国、巴基斯坦、印度、马来西亚、印度尼西亚、菲律宾等地。

## 生长习性

喜温喜光，适应性强，冬季温度不能低于10℃，夏季高温时要稍加遮阴；对水分要求不高。

## 用途、价值

可食用，也可作香料。色彩鲜艳、品种甚多，是应用较广的观叶花卉。可配置图案花坛，还可作为花篮、花束的配叶使用。

## PLANT MORPHOLOGY

Erect or ascending herb. Stem purple, quadrangular, puberulent, branched. Leaves ovate, 4-13cm in length, crenate, yellow, crimson, purple and green, puberulent on both sides, and sparsely covered with reddish brown glandular dots below. Cymes form panicle; calyx campanulate, 2-3mm in length; corolla purple or blue, 2cm in length, pilose; flowering period in July. Nutlets, broadly ovoid or rounded, compressed, brown, glossy, 1-2mm in length.

## ECOLOGICAL DISTRIBUTION

Native to China. Distributed in China, Pakistan, India, Malaysia, Indonesia, Philippines, etc.

## GROWTH HABIT

Wet and light-favored, strong adaptability, the temperature in winter cannot be less than 10℃, shaded slightly in high temperature in summer; not strict to the moisture quality.

## APPLICATION,VALUE

Edible or used as perfume. It is a widely used foliage flower because of its bright color and many varieties. It can be configured with pattern flower bed, and used as the leaf of flower basket and flower bundle.

# 57 / 假马鞭

*Stachytarpheta jamaicensis*

马鞭草科 Verbenaceae　假马鞭属 *Stachytarpheta*

别名：蛇尾草、蓝草、大种马鞭草、玉龙鞭

## 植物形态

多年生粗壮草本。高0.6~2m。幼枝近四方形，疏生短毛。叶片厚纸质，长2~8cm，边缘有粗锯齿。穗状花序顶生，长11~29cm；花单生于苞腋；一半嵌生于花序轴的凹穴中，螺旋状着生；花萼管状，膜质；花冠深蓝紫色，长1~2cm，内部有毛；花期8月。果内藏于膜质的花萼内，成熟后2瓣裂，每瓣有1枚种子；果期9~12月。

## 生态分布

原产中南美洲。分布于中国、巴基斯坦、巴西、阿根廷、越南等地。

## PLANT MORPHOLOGY

Perennial robust herb. Plant height 0.6-2m. Young branches nearly square, sparsely short hairy. Leaf blade thick papery, 2-8cm in length, margin with coarse serration. Spikes terminal, 11-29cm in length; flowers solitary in axils; half embedded in the pit of inflorescence axis, spirally inserted; calyx tubular, membranous; corolla dark blue purple, 10-20mm in length, hairy inside; flowering period in August. Fruit stored in the membranous calyx, with 2 petals split after maturity, each petal 1 seed; fruiting period from September to December.

## ECOLOGICAL DISTRIBUTION

Native to central and South America. Distributed in China, Pakistan, Brazil, Argentina, Vietnam, etc.

## 生长习性

喜温暖湿润的环境，适生温度 22~32℃；生命力强，耐热、耐旱、耐瘠薄。

## 用途、价值

叶形优美，具观赏价值。全草可入药，可作发汗药，有凉血散瘀、清热消毒等功效。

## GROWTH HABIT

Favoring warm and humid environment, Suitable for temperature is 22-32℃, strong vitality, heat resistance, drought tolerance and barren tolerance.

## APPLICATION,VALUE

Leaves are beautiful and have ornamental value. Used as medicine and sweating medicine. Effective in cooling blood, dispersing blood stasis, clearing heat and disinfecting.

# 58 / 曼陀罗

*Datura stramonium*
茄科 Solanaceae  曼陀罗属 *Datura*
别名：土野麻子、洋金花、万桃花、狗核桃

## 植物形态

草本。高0.5~1.5m。植株无毛或被短柔毛，茎粗壮，圆柱状，淡绿色或带紫色。叶广卵形，边缘有不规则波状浅裂，叶柄长3~5cm。花单生叶腋，直立，有短梗；花萼筒状，长4~5cm，基部稍膨大，顶端紧围花冠筒，宿存部分随果实增大并向外反折；花冠漏斗状；花期6~10月。蒴果，直立，长3~5cm，有时被坚硬针刺，淡黄色；果期7~11月。种子卵圆形；黑色。

## 生态分布

原产印度。分布于中国、巴基斯坦等地。

## 生长习性

喜温暖、阳光充足的环境，适生于排水良好的砂质壤土。多生于田间、沟旁、道边、河岸、山坡等。

## 用途、价值

花艳丽妖娆，有较高的观赏价值。种子油可制肥皂和掺和油漆；叶、花、籽均可入药，用于麻醉；花能祛风湿、止喘定痛，亦可治神经痛等。

## PLANT MORPHOLOGY

Herb. Plant height 50-150cm. Plants glabrous or puberulent, stems thick, cylindrical, light green or purple. Leaves broadly ovate, with irregular wavy lobes on the edge, and petiole 3-5cm in length. Flowers solitary leaf axils, erect, with short stalks; calyx tube-shaped, 4-5cm in length, slightly enlarged at base, closely surrounding corolla tube at the top, increase of persistent part corresponding to fruit growth, folding outward; corolla funnel-shaped; flowering period from June to October. Capsule, erect, 3-5cm in length, sometimes pricked with hard needles, light yellow; fruiting period from July to November. Seeds ovoid; black.

## ECOLOGICAL DISTRIBUTION

Native to India. Distributed in China, Pakistan, etc.

## GROWTH HABIT

Favoring warm and sunshine environment, favored growing in sandy soil with good drainage. Common in fields, ditches, roadsides, river banks, hillsides, etc.

## APPLICATION,VALUE

The flowers are gorgeous and enchanting, with high ornamental value. Seed oil can be used to make soap and blend with paint; leaves, flowers, and seeds used as medicine for anesthesia; flowers can relieve rheumatism, relieve asthma and relieve pain, and can treat neuralgia.

# 59 / 蝴蝶兰

## 植物形态

　　草本。茎短，常被叶鞘所包。叶片稍肉质，背面紫色，椭圆形、长圆形或镰刀状长圆形，长10~20cm，宽3~6cm。花序侧生于茎基部，长达50cm；花序柄绿色，被数枚鳞片状鞘；花序轴紫绿色，常具数朵由基部向顶端逐朵开放的花；花白色，花瓣菱状圆形，长2.7~3.4cm，宽2.4~3.8cm，先端圆形，基部收狭呈短爪，具网状脉，唇瓣3裂，基部具长爪。花期4~6月。

## 生态分布

　　原产马来西亚。分布于中国、巴基

## PLANT MORPHOLOGY

　　Herb. Stem short, often covered by leaf sheaths. Leaves slightly fleshy, purple on the back, oval, oblong or sickle-shaped oblong, 10-20cm in length and 3-6cm in width. Inflorescence located laterally at the base of the stem, up to 50cm in length; e inflorescence with a green stalk and covered with several scaly sheaths; inflorescence axis purple-green, often with several flowers that open one by one from the base to the top; flowers white with rhomboid petals, 27-34mm in length , 24-38mm in width, apex round, narrow at the base and short claws, with reticulated veins, lip 3-lobed, and long claws at the base. Flowering period from April to June.

斯坦、泰国、菲律宾、印度尼西亚
等地。

## 生长习性

生于热带雨林地区，喜暖畏寒。适
生温度15~20℃。附生生长于高温高湿
河川、海岸边的树木上。

## 用途、价值

观赏价值高，可作为盆花和鲜切花
销售；能吸收空气中的养分而生存；色
彩丰富，对原种进行人工杂交，可改良
出各种花色、花型。

Native to Malaysia. Distributed in China,
Pakistan, Thailand, Philippines, Indonesia, etc.

### GROWTH HABIT

Growing in the tropical rain forest area and
favoring warm, but chilly. Suitable for temperature
is 15-20°C. Epiphytic forest trees growing along the
coast of high-temperature and high-humidity rivers.

### APPLICATION,VALUE

High ornamental value and can be sold as
potted flowers and fresh cut flowers; it can absorb
nutrients in the air and survive; it has a variety of
colors and can be artificially mated with the original
species to improve various flower colors and flower
types.

# 60 / 文殊兰

*Crinum asiaticum*

石蒜科 Amaryllidaceae　文殊兰属 *Crinum*
别名：罗裙带

## 植物形态

多年生粗壮草本。鳞茎长圆柱形。叶深绿色，线状披针形，长0.5~1m，宽7~12cm，边缘波状。伞形花序，佛焰苞状总苞片披针形，膜质，花高脚碟状；花被管纤细，伸直，绿白色，花被白色；花期夏季。蒴果，近球形，直径3~5cm。

## 生态分布

原产印度尼西亚苏门答腊等地。分布于中国、巴基斯坦、缅甸、菲律宾等地。

## PLANT MORPHOLOGY

Stout perennial herb. Bulb long and cylindrical. Leaves dark green, linear-lanceolate, 50-100cm in length and 7-12cm in width, with wavy edges. Umbrella, bud-like involucral bracts lanceolate, membranous, flower tall and dish-shaped; perianth tube slender, straight, green and white, perianth white; flowering in summer. Capsule, nearly spherical, 3-5cm in diameter.

## ECOLOGICAL DISTRIBUTION

Native to Indonesia, Sumatra, etc. Distributed in China, Pakistan, Myanmar, Philippines, etc.

## GROWTH HABIT

Favoring warm. Suitable for temperature is 15-20°C. Not cold-resistant, saline-alkali resistance,

## 生长习性

喜温暖。适生温度15~20℃。不耐寒，耐盐碱，盆栽土以腐殖质含量高、疏松肥沃、通透性强的砂质培养土为宜。

## 用途、价值

花叶美观，观赏价值高，可作园林景区、校园等草坪的点缀品，是一种庭院装饰花卉；盆栽可置于会议厅、宾馆、宴会厅门口等。叶与鳞茎可药用，有活血散瘀、消肿止痛等功效；全株有毒。

potting soil should be sandy culture soil with high humus, loose and fertile, and strong permeability.

### APPLICATION,VALUE

The flowers and leaves are beautiful and have high ornamental value. Used as ornaments of lawn in garden scenic spots and campus and a kind of garden decorative flowers. Potted plants can be placed at the entrance of conference hall, hotel and banquet hall. The leaves and bulbs can be used medicinally, and effective in promoting blood circulation, dispersing blood stasis, detumescence and pain relief; the whole plant is poisonous.

# 61 / 龙舌兰

*Agave americana*

天门冬科 Asparagaceae　龙舌兰属 *Agave*

别名：金边龙舌兰、龙舌掌、番麻

## 植物形态

多年生草本。叶基生，呈莲座状，肉质，倒披针形，长1~2m，宽15~20cm，先端具暗褐色硬尖刺，叶缘疏生刺状小齿。花茎粗壮，高达6m，圆锥花序，花黄绿色，花被筒长约1.2cm；开花后花序生出少数珠芽。蒴果，长圆形，长约5cm。

## 生态分布

原产美洲热带。中国、巴基斯坦等地常引种栽培。

## PLANT MORPHOLOGY

Perennial herb. Leaves rosette-like, fleshy, oblanceolate, 1-2m in length and 15-20cm in width, with dark brown hard spines at the apex, and sparsely spiny denticles on the leaf margin. Flower stem thick, up to 6m in length, panicle, yellow green flowers, perianth tube approximately 1.2cm in length; after flowering, inflorescence producing a few bulbils. Capsule, oblong, about 5cm in length.

## ECOLOGICAL DISTRIBUTION

Native to tropical America, often introduced and cultivated in China, Pakistan, etc.

## GROWTH HABIT

Light-favored, not tolerant of shade, strong drought resistance, slightly hardy, suitable for

## 生长习性

喜光，不耐阴，耐旱力强，稍耐寒，适生温度15~25℃；适生于冬季冷凉干燥环境。对土壤要求不严，适生于疏松、肥沃及排水良好的湿润砂质土壤。

## 用途、价值

是一种室内观赏植物。茎、叶基部柔软的分生组织可食用；叶可作饲料；还可用于造纸、生物燃料。

15-25°C temperature; suitable for cold and dry environment in winter. Lax requirements on the soil, suitable for moist sandy soil with loose, fertile and good drainage.

### APPLICATION,VALUE

Indoor ornamental plant. Soft meristem at the base of stem and leaf edible; leaf applicable in feed; also applicable in papermaking and biofuel.

# 62 / 虎尾兰

*Sansevieria trifasciata*
龙舌兰科 Agavaceae　虎尾兰属 *Sansevieria*
别名：金边虎尾兰

## 植物形态

　　多年生草本。叶基生，直立，硬革质，扁平，长条状披针形，长30~70cm，宽3~5cm，有白绿色和深绿色相间的横带斑纹，边缘绿色，向下部渐狭成柄。花葶高30~80cm，基部有淡褐色的膜质鞘；花淡绿色或白色，簇生，排成总状花序，花梗长5~8mm，花期11~12月。浆果，直径7~8mm。

## 生态分布

　　原产非洲西部及亚洲南部。分布于中国、印度、巴基斯坦等地。

## PLANT MORPHOLOGY

　　Perennial herb. Leaves basal, upright, hard leather, flat, long lanceolate, 30-70cm in length, 3-5cm in width, with white-green, indistinct, and glorious-green horizontal stripes, with green edges and tapering downwards. handle. Scape 30-80cm in height, with a light brown membranous sheath at the base; flowers light green or white, clustered, arranged in racemes, pedicels 5-8mm in length, flowering period from November to December. Berries, about 7-8mm in diameter.

## ECOLOGICAL DISTRIBUTION

　　Native to western Africa and southern Asia. Distributed in China, India, Pakistan, etc.

## 生长习性

喜光，耐旱，适生温度20~30℃。适生于疏松、透气性好的土壤。

## 用途、价值

对环境适应能力强，栽培利用广泛，为常见的盆栽观叶植物；可用于净化空气，吸收室内部分有害气体。叶可药用，有清热解毒、活血消肿等功效。

## GROWTH HABIT

Thermophilic, drought resistance, Suitable for 20-30°C temperature. Suitable for loose soil with good air permeability.

## APPLICATION,VALUE

Strong adaptability to the environment and widely used for cultivation. common potted foliage plant in the courtyard; applicable in purify the air and absorb some harmful indoor gases. Leaves applicable in medicine to clear away heat and toxins; promote blood circulation and reduce swelling.

# 63 / 一串红

## 植物形态

亚灌木状草本。高5~90cm。叶卵圆形或三角状卵圆形，稀钝，边缘具锯齿；叶柄长3~5cm。轮伞花序具2~6花，组成总状花序；苞片卵形，红色，花前包被花蕾；花梗密被红色腺柔毛；花萼红色，钟形；花冠鲜红色，被柔毛。小坚果，暗褐色，顶端不规则皱褶，边缘具窄翅；花期3~10月。

## 生态分布

原产巴西。分布于中国、巴基斯坦等地。

## 生长习性

耐寒性差，喜光，也耐半阴。适生温度20~25℃。适生于疏松、肥沃、排水良好的砂质壤土。

## 用途、价值

观赏价值高，常用作花丛、花坛的主体材料；也可植于带状花坛或自然式栽植于林缘，与浅黄色美人蕉、矮万寿菊等配合布置。

## PLANT MORPHOLOGY

Subshrub herb. Plant height 5-90cm. Leaves ovoid or triangular-ovate, blunt, with serrated edges; petiolese 3-5cm in length. Verticillaster with 2-6 flowers, composed of racemes; bracts ovate, red, covered with flower buds; peduncle densely red glandular pilose; calyx red, bell-shaped; corolla bright red, pilose; small nuts, dark brown, irregularly creped at the top, with narrow wings on the edge; flowering period from March to October.

## ECOLOGICAL DISTRIBUTION

Native to Brazil. Distributed in China, Pakistan, etc.

## GROWTH HABIT

It has poor cold resistance, likes light, but also bears half shade. The suitable temperature for growth is 20-25°C. It is suitable for loose, fertile and well drained sandy loam.

## APPLICATION,VALUE

It has high ornamental value and is often used as the main material of flower beds; it can also be planted in belt-shaped flower beds or planted in forest edges; it can be arranged with light yellow canna and dwarf marigolds.

# 64 / 奔龙花

番杏科 Aizoaceae　　旭峰花属 *Cephalophyllum*

*Cephalophyllum framesii*

## 植物形态

多年生草本。叶较长，绿色，肉质，易折断。是一种常见的多肉植物。

## 生态分布

原产非洲南部。分布于中国、巴基斯坦、马达加斯加等地。

## 生长习性

喜温暖干燥且阳光充足的环境，适生温度15~30℃。花常在中午阳光下开放，傍晚时会自动闭合，待次日再开，可连续开3~5天，雏菊型。适生于排水良好的肥沃土壤。

## PLANT MORPHOLOGY

Perennial herb. Leaves long, green, fleshy and easy to break. common succulent plant.

## ECOLOGICAL DISTRIBUTION

Native to southern Africa. Distributed in China, Pakistan, Madagascar, etc.

## GROWTH HABIT

Thermophilic, dry and sunny environment. suitable for 15-30°C temperature. Flowers usually open in the sun at noon, and close automatically in the evening, and open again the next day. Opening for 3-5 days continuously. Suitable for fertile soil with good drainage.

## 用途、价值

　　可用于园林栽培观赏。植株形态优美，花色艳丽；可布置于庭院、大堂、客厅等。

## APPLICATION,VALUE

　　Applicable in garden cultivation and ornamental. beautiful in shape and gorgeous in color. arranged in courtyard, lobby, living room, etc.

# 65 / 美人蕉

*Canna indica*

美人蕉科 Cannaceae　　美人蕉属 *Canna*

别名：蕉芋

## 植物形态

多年生草本。高50~150cm。全株绿色，无毛，被蜡质白粉。具块状根茎。叶互生，卵状长圆形，长10~30cm，宽8~15cm。总状花序，疏花；花红色，单生；苞片卵形，绿色；花冠裂片披针形，绿或红色。蒴果，长卵形，绿色，有软刺，长1~2cm；花果期3~12月。

## 生态分布

原产印度。分布于中国、巴基斯坦、印度等地。

## 生长习性

喜温暖湿润气候，不耐霜冻，适生温度25~30℃。对土壤要求不严，耐瘠薄，适生于肥沃、湿润、排水良好的土壤。

## 用途、价值

花大色艳，观赏价值高，具有美化环境、净化空气的作用。茎叶纤维可制成人造棉、麻袋、绳子等；叶可提取芳香油，残渣用于造纸；根茎可治黄疸型急性传染性肝炎。

## PLANT MORPHOLOGY

Perennial herb. Plant height 50-150cm. Green, glabrous and covered with waxy white powder. With massive rhizomes. Leaves alternate, ovate oblong, 10-30cm in length and 8-15cm in width. Racemes, sparsely flowered; flowers red, solitary; bracts ovate, green; corolla lobes lanceolate, green or red. Capsule, long ovate, green, spiny, 1-2cm in length; flowering and fruiting period from March to December.

## ECOLOGICAL DISTRIBUTION

Native to India. Distributed in China, Pakistan, India, etc.

## GROWTH HABIT

Favoring warm and humid climate, not frost-resistant. Suitable for 25-30℃ temperature. not strict to the soil quality, barren resistance. suitable for fertile, moist and well drained soil.

## APPLICATION,VALUE

Flower big and colorful, with high ornamental value. effective in beautify the environment and purify the air. The fiber of stem and leaf applicable in cotton, sack, rope etc; the aromatic oil can be extracted from leaves, and the residue used for papermaking; the rhizome used to treat icteric acute infectious hepatitis.

## 66 / 花烛

*Anthurium andraeanum*
天南星科 Araceae　花烛属 *Anthurium*
别名：红鹅掌、火鹤花、安祖花、红掌

## 植物形态

多年生常绿草本。茎节短。叶自基部生出，革质，全缘，长圆状心形或卵心形，叶柄细长。佛焰苞平出，卵心形，革质，并有蜡质光泽，橙红色或猩红色；肉穗花序，长5~7cm，黄色，可常年开花不断。

## 生态分布

原产哥斯达黎加、哥伦比亚等热带雨林区。分布于中国、巴基斯坦、安哥拉、赞比亚、加拿大、巴西、阿根廷等地。

## PLANT MORPHOLOGY

Perennial evergreen herb. Stem node short. Leaves from the base, leathery, entire, oblong heart-shaped or ovate heart-shaped, petiole slender. Spathe flat, ovate heart-shaped, leathery, with waxy luster, orange red or scarlet; fleshy inflorescence, 5-7cm in length, yellow, bloom all year round.

## ECOLOGICAL DISTRIBUTION

Native to tropical rain forest areas such as Costa Rica and Colombia. Distributed in China, Pakistan, Angola, Zambia, Canada, Brazil, Argentina, etc.

## GROWTH HABIT

Favoring warm, hot and humid semi shade environment, not drought and strong sunlight resistance. suitable for 20-32°C temperature.

## 生长习性

喜温热多湿的半阴环境，不耐旱和强光暴晒，适生温度20~32℃。

## 用途、价值

花叶具观赏价值，花期长，为优质盆栽和切花材料。可吸收人体排出的废气，也可吸收装修残留的各种有害气体等，同时可保持空气湿润。

## APPLICATION,VALUE

Flowers and leaves ornamental value and long flowering period. high-quality potted and cut flower materials. absorb the exhaust gas discharged from human body and various harmful gases left over from decoration, and keep the air moist at the same time.

# 67 / 向日葵

菊科 Asteraceae　向日葵属 *Helianthus*

*Helianthus annuus*

别名：朝阳花

## 植物形态

　　一年生高大草本。高1~3m。茎直立，粗壮，被白色粗硬毛。叶互生，心状卵圆形，基出三出脉，边缘有粗锯齿，两面被短糙毛。头状花序，极大，径10~30cm，单生，常下倾；花托平或稍凸，有半膜质托片；舌状花多数，黄色，舌片开展，长圆形；管状花多数，棕色或紫色，有披针形裂片；花期7~9月。瘦果，倒卵形或卵状长圆形，稍扁压，常被白色短柔毛；果期8~9月。

## PLANT MORPHOLOGY

　　Annual tall herb. Plant height 1-3m. Stems erect, stout, with white thick hairs. Leaves alternate, heart-shaped oval, base out of three veins, margin coarsely serrate, and both surfaces with short scabrous. Inflorescences capitate, extremely large, about 10-30cm in diameter, solitary and often downdip; receptacle flat or slightly convex, with semi membranous stipules; ligulate flowers mostly yellow, tongue s spreading and oblong. Most tubular flowers, brown or purple, with lanceolate lobes; flowering period from July to September. Achenes, obovate or ovate oblong, slightly compressed, often white pubescent; fruiting period from August to September.

## 生态分布

原产北美。分布于中国、巴基斯坦、俄罗斯、秘鲁、埃及、阿根廷等地。

## 生长习性

喜温，耐寒，抗旱，耐盐碱。适应性较强，种子耐低温能力强。短日照作物，向光性强。对水分需求较多。对土壤要求不严。

## 用途、价值

种子可食用，葵花籽营养丰富，种子可榨成食用油。脱壳的葵花籽仁可烹饪或用于制作蛋糕、冰淇淋等甜食。有较高的观赏价值。全株可入药，种子有驱虫止痢、降脂的作用；花盘可清热化痰、凉血止血；茎髓为利尿消炎剂；叶与花瓣可清热解毒，还可作健胃剂；花穗、种壳及茎秆可作饲料及工业原料，可制人造丝及纸浆等。对金属污染物有较强的抵御能力，根部的富集作用可吸收有害污染物。

## ECOLOGICAL DISTRIBUTION

Native to North America. Distributed in China, Pakistan, Russia, Peru, Egypt, Argentina, etc.

## GROWTH HABIT

Favoring warm, cold resistance, drought resistance, salt and alkali resistance. It has strong adaptability. The low temperature tolerance of seeds is strong. Short day crops have strong phototropism. It needs more water. The soil is not strict.

## APPLICATION,VALUE

The seeds are edible, sunflower seeds are rich in nutrition, and the seeds can be squeezed into edible oil. Shelled sunflower seeds can be cooked or used to make cakes, ice cream and other sweets. It has high ornamental value. The whole plant can be used as medicine, the seeds have the effect of expelling insects, stopping dysentery and reducing blood lipid; the flower plate can clear away heat and phlegm, cool blood and stop bleeding; the stem pith is a diuretic and anti-inflammatory agent; the leaves and petals can clear away heat and detoxify, and can also be used as a stomach tonic; the flower spike, seed shell and stem can be used as feed and industrial raw materials, as well as rayon and pulp. It has strong resistance to metal pollutants, and the accumulation of roots can absorb harmful pollutants.

# 68 / 石竹

*Dianthus chinensis*
石竹科 Caryophyllaceae　石竹属 *Dianthus*
别名：长萼石竹、丝叶石竹、蒙古石竹、北石竹、山竹子

## 植物形态

多年生草本。高30~50cm。全株无毛，带粉绿色；茎由根颈生出，疏丛生，直立，上部分枝。叶线状披针形，长3~5cm，宽2~4mm，全缘或具微齿。花单生枝端或数花集成聚伞花序；花梗长10~30mm，花萼筒形，具纵纹；花瓣倒卵状三角形，紫红、粉红、鲜红或白色，先端不整齐齿裂，花期5~6月。蒴果，圆筒形，包于宿存萼内；果期7~9月。种子扁圆形。

## 生态分布

原产中国。分布于巴基斯坦、俄罗斯、朝鲜等地。

## 生长习性

喜阳光充足、干燥、通风、凉爽湿润的环境。耐寒。适生于肥沃、疏松含石灰质的壤土或砂质壤土。

## 用途、价值

观赏花卉；园林中可用于花坛、花境等，也可用于岩石园和草坪边缘点缀；大面积成片栽植可作景观地被材料。可吸收二氧化硫和氯气等有害气体。全草可入药，有清热利尿、散瘀消肿等功效。

## PLANT MORPHOLOGY

Perennial herb. Plant height 30-50cm. Whole plant glabrous and pinkish green; stem growing from the root neck, sparsely clumped, erect, and branched above. Leaves linear-lanceolate, 3-5cm in length, 2-4mm in width, fully marginal or micro dentate. Flowers solitary branch at ends or several flowers integrated into umbrella inflorescences; peduncles 10-30mm in length, calyx tubular, longitudinal; petals inverted ovate triangular, purple-red, pink, bright red or white, with irregular teeth at the apex, and flowering period from May to June. Capsule, cylindrical, enclosed in a calyx; fruiting period from July to September. Seeds flattened and rounded.

## ECOLOGICAL DISTRIBUTION

Native to China. Distributed in Pakistan, Russia, North Korea, etc.

## GROWTH HABIT

Light-favored, dry, ventilated, cool and humid environment. Cold resistant. It is suitable for loam or sandy loam with rich and loose calcareous content.

## APPLICATION,VALUE

Ornamental flowers; garden can be used for flower beds, flower border, can also be used for rock garden and lawn edge decoration; large area of planting can be used as landscape ground cover materials. It can absorb harmful gases such as sulfur dioxide and chlorine. The whole herb can be used as medicine and has the effects of clearing heat and diuresis, dispersing blood stasis and detumescence.

# 69 / 黄秋英

*Cosmos bipinnatus*
菊科 Asteraceae　秋英属 *Cosmos*
别名：波斯菊、大波斯菊、秋樱、格桑花、扫地梅

## 植物形态

一年生或多年生草本。高1~2m。茎无毛或稍被柔毛。二回羽状复叶，叶对生，深裂，叶缘粗糙，裂片针形。头状花序单生，径3~6cm，花序梗长6~18cm；总苞片淡绿色，具深紫色条纹；舌状花紫红、粉红或白色；管状花黄色；花期6~8月。瘦果，黑紫色，上端具长喙，有2~3尖刺；果期9~10月。

## 生态分布

原产墨西哥。分布于中国、巴基斯坦、巴西、印度、日本、朝鲜、韩国等地。

## 生长习性

喜温暖，不耐寒，忌干旱。适生于肥沃、疏松和排水良好的微酸性砂质壤土。

## 用途、价值

可供观赏，多株丛植或片植；也可用于花境栽植等。全草可入药，具有清热解毒等功效，可辅助治疗急性、慢性细菌性痢疾等症。

## PLANT MORPHOLOGY

Annual or perennial herb. Plant height 1-2m. Stem glabrous or slightly pilose. leaves bipinnately compound, opposite, deeply lobed, with rough leaf margin and needle like lobes. Capitate inflorescence, solitary, 3-6cm in diameter, 6-18cm in peduncle length; bracts pale green with dark purple stripes; tongue flowers purple, pink or white; tubular flowers yellow; the flowering period from June to August. Achene, black purple, with long beak at the upper end, with 2-3 spines; fruiting period from September to October.

## ECOLOGICAL DISTRIBUTION

Native to Mexico. Distributed in China, Pakistan, Brazil, India, Japan, North Korea, South Korea, etc.

## GROWTH HABIT

Favoring warm. It is not resistant to cold and drought. It is suitable for loam with rich, loose and good drainage.

## APPLICATION,VALUE

It can be used for ornamental, multi plant cluster planting or piece planting, and can also be used for flower border planting. The whole herb can be used as medicine, which has the effect of clearing away heat and detoxification, and can be used to treat acute, chronic and bacillary dysentery.

# 70 / 紫苜蓿

豆科 Leguminosae

*Medicago sativa*
苜蓿属 *Medicago*
别名：苜蓿

## 植物形态

多年生草本。高30~100cm。茎四棱形。羽状三出复叶；托叶大，基部全缘或具1~2齿裂；叶上缘具锯齿。总状或头状花序；花梗挺直；花萼被伏柔毛；花冠多色：淡黄、深蓝至暗紫色；花瓣均具长瓣柄；花期5~7月。荚果，螺旋状，被柔毛或渐脱落，熟时棕色；果期6~8月。种子10~20枚，卵形，平滑，黄色或棕色。

## 生态分布

原产小亚细亚及伊朗地区。广泛分布于中国、巴基斯坦、巴西、埃及、阿

## PLANT MORPHOLOGY

Perennial herb. Plant height 30-100cm. Stem quadrilateral. Pinnately, triple compound leaves. Stipules large, base entire or with 1-2 teeth cleft; leaf margin serrate. Racemose or capitate inflorescence; pedicel straight; calyx covered with soft hair; corolla polychromatic: light yellow, dark blue to dark purple; petals with long petiole, flowering period from May to July. Legume, heliciform, pilose or glabrescent, brown when ripe; fruiting period from June to August. Seeds 10-20, ovate, smooth, yellow or brown.

## ECOLOGICAL DISTRIBUTION

Native to Asia Minor and Iran. Widely distributed in China, Pakistan, Brazil, Egypt, Argentina, Russia, etc.

根廷、俄罗斯等地。

## 生长习性

喜温，耐旱，耐寒，适应性强，可生长于-30℃的极端低温。根系发达。

## 用途、价值

适应性强，栽种范围广，再生能力强，富含膳食纤维、矿物质、维生素、粗蛋白质等，是一种极为优良的饲料作物，有"牧草之王"之称。根系发达，固氮，能提高土壤有机质的含量，防风固沙，减少地表径流，为优良的水土保持植物。

## GROWTH HABIT

Favoring warmth. It is resistant to drought and cold. It can grow at the extreme temperature of -30°C. The root system is well developed.

## APPLICATION,VALUE

It has strong adaptability, wide planting range, strong regeneration ability, rich in dietary fiber, minerals, vitamins, crude protein, etc. it is an excellent feed crop, known as "the king of forage". The root system is well developed. Nitrogen fixation can increase the content of soil organic matter, prevent wind and sand, reduce surface runoff, and is an excellent plant for soil and water conservation.

# 71 / 绿萝

*Epipremnum aureum*

天南星科 Araceae 麒麟叶属 *Epipremnum*

别名：小绿、魔鬼藤、黄金葛、黄金藤

## 植物形态

常绿草本。茎攀缘，节间具纵槽；多分枝，枝悬垂；节间长15~20cm。叶片翠绿色，有时具不规则的黄色斑块，基部深心形；叶柄两侧具鞘，鞘革质，宿存，下部叶片大，长5~10cm，上部的长6~8cm；成熟枝上叶柄粗壮，基部稍扩大，上部关节稍肥厚。

## 生态分布

原产所罗门群岛。分布于中国、巴基斯坦、日本、韩国、埃及等地。

## PLANT MORPHOLOGY

Evergreen herb. Stems climbing, internodes with longitudinal grooves; much branched, branches hanging; internodes 15-20cm in length. Leaf blade emerald green with irregular yellow plaques and deep heart-shaped at the base; both sides of petiole sheathed, leathery and persistent; lower leaf blade large, 5-10cm in length, and the upper part 6-8cm in length; on the mature branch, petiole stout, base slightly enlarged, and upper joint slightly thickened.

## ECOLOGICAL DISTRIBUTION

Native to Solomon Islands. Distributed in China, Pakistan, Japan, South Korea, Egypt, etc.

## GROWTH HABIT

Favoring humid environment, avoid direct

## 生长习性

喜潮湿的环境，忌阳光直射。越冬温度不应低于15℃。生命力强。适生于富含腐殖质、疏松肥沃、微酸性的土壤。

## 用途、价值

观赏价值高，缠绕性强，四季常绿，是优良的观叶植物。可攀附，用于门厅、宾馆装饰或置于书房、窗台等，亦可作为地被植物。具有极强的空气净化功能。

sunlight. The overwintering temperature should not be lower than 15°C. Strong vitality. It is suitable for the soil rich in humus, loose and fertile, and slightly acidic.

**APPLICATION,VALUE**

It is evergreen in all seasons and has high ornamental value. It is an excellent foliage plant. It can be attached, used for decoration of hallways and hotels, or placed in study, windowsill, etc., and can also be used as ground cover plants. It has strong air purification function.

# 72 / 墨苜蓿

茜草科 Rubiaceae　墨苜蓿属 *Richardia*

*Richardia scabra*

## 植物形态

一年生匍匐或近直立草本。高10~80cm。茎近圆柱形，被硬毛，节上无不定根，疏分枝。叶厚纸质，卵形、椭圆形或披针形，长1~5cm；托叶鞘状，顶部截平。头状花序，顶生，总梗顶端有1~2对叶状总苞，花冠白色，漏斗状或高脚碟状，花期春夏间。分果瓣3~6，长2~3.5mm，长圆形至倒卵形，背部密覆小乳突和糙伏毛。

## 生态分布

原产热带美洲。分布于中国、巴基斯坦、巴西、阿根廷等地。

## PLANT MORPHOLOGY

Annual creeping or nearly erect herb. Plant height 10-80cm. Stem subcylindrical, hirsute, without adventitious roots on nodes, sparsely branched. Leaves thick papery, ovate, elliptic or lanceolate, 1-5cm in length; stipules sheathed, apically truncate. Inflorescence capitate, terminal, with 1-2 pairs of leafy involucres at the top of the total stem. Corolla white, funnel-shaped or high foot saucer shaped. Flowering period between spring and summer. Fruit petals 3-6, 2-3.5mm in length, oblong to obovate, back densely covered with small milky convex and strigose.

## ECOLOGICAL DISTRIBUTION

Native to tropical America. Distributed in China, Pakistan, Brazil, Argentina, etc.

## 生长习性

喜高温多湿的生长环境，也耐旱，耐盐碱，对土壤要求不严。

## 用途、价值

可作为荒漠地区防风固沙的地被植物，可在海岸流动沙丘上生长。根和种子可入药，具有催吐、利尿等功效。

## GROWTH HABIT

Favoring hot and humid environment. It is resistant to drought and salt. The soil is not strict.

## APPLICATION,VALUE

It can be used as a ground cover plant for windbreak and sand fixation in desert areas, and can grow on the moving sand dunes along the coast. Root and seed can be used as medicine, with emetic, diuretic and other effects.

# 73 / 铺地黍

## 植物形态

多年生草本。高50~100cm。根茎粗壮发达。叶鞘光滑，边缘被纤毛；叶舌顶端被毛。叶片质硬，线形，长5~25cm，宽2.5~5mm，干时常内卷，呈锥形，上表皮粗糙或被毛。圆锥花序开展，长5~20cm，分枝具棱槽；小穗长圆形，长约3mm。花果期6~11月。

## 生态分布

原产中国。现分布于巴基斯坦等地。

## 生长习性

对生长环境条件要求不严，多生于路边、山坡草地和近海沙地上，也生长于旱作物地中。

## 用途、价值

是巴基斯坦地区入侵植物。旱地的一种地区性恶性杂草，生长迅速，粗大根茎深入土层，能刺穿作物根部，抢夺田间大量肥分；地上部分则遮盖作物茎叶，使田间通风透光不良，从而影响作物的生长发育。是稻纵卷叶螟的寄主，并感染作物的瘟病、锈病、黑粉病等。该种也是草坪的主要害草之一，潜在危害大，难根除，目前尚无理想的除草剂可防治，应避免侵入。

## PLANT MORPHOLOGY

Perennial herb. Plant height 50-100cm. Rhizomes stout and developed. Leaf sheaths smooth, margin ciliated; ligule tip hairy. Leaf blade hard, linear, 5-25cm in length, 2.5-5mm in width, often involute when dry, tapered, upper epidermis rough or hairy. Panicle spreading, 5-20cm in length, branches ribbed; spikelet oblong, about 3mm in length. Flowering and fruiting period from June to November.

## ECOLOGICAL DISTRIBUTION

Native to China. Distributed in Pakistan, etc.

## GROWTH HABIT

The requirements for the growth environment are not strict; it grows mostly in the roadside, hillside grassland and coastal sandy land, but also in the dry crop fields.

## APPLICATION, VALUE

It is an invasive plant in Pakistan. It is a kind of regional malignant weed in dry land. It grows rapidly, and its thick rhizome goes deep into the soil layer, which can pierce the root of crops and seize a large amount of fertilizer in the field. The aboveground part covers the stems and leaves of crops, making the field ventilation and light transmission poor, thus affecting the growth and development of crops. It is the host of *Cnaphalocrocis medinalis*, and may be infected with crop blast, rust and smut. This species is also one of the main harmful weeds in lawn, which has high damage potential and is difficult to eradicate. At present, there is no ideal herbicide to control, so the invasion should be avoided.

# 74 / 巨菌草

*Pennisetum giganteum*
禾本科 Poaceae　狼尾草属 *Pennisetum*

## 植物形态

在温度适宜地区为多年生植物。植株高大，直立、丛生，根系发达。在福建省生长半年，茎粗可达3.5cm；节间长9~15cm；15个有效的分蘖，每节着生1个腋芽，并由叶片包裹；叶片互生，长60~130cm，宽3.5~6cm。

## 生态分布

原产地在北非。分布于巴基斯坦、非洲，引种于中国等地。

## 生长习性

其光合与蒸腾之比较低，因此，其

## PLANT MORPHOLOGY

A perennial plant when in areas with suitable temperature. Plant tall, erect, clustered, and developed root system. It grows in Fujian Province in China for half a year, stem thickness can reach 3.5cm; internodes 9-15cm in length; 15 effective tillers, each node has an axillary bud, surrounded by leaves; leaves alternate, 60-130cm in length, 3.5-6cm in width.

## ECOLOGICAL DISTRIBUTION

Native to North Africa. Distributed in Pakistan, Africa, and introduced in China, etc.

## GROWTH HABIT

Its photosynthesis and transpiration ratio is low, therefore, its growth requires moist soil conditions

生长除需高温外，需湿润的土壤条件，能耐短期的干旱，但不耐涝。

## 用途、价值

抗逆性强，产量高，粗蛋白和糖分含量高，是高产优质的菌草之一，用巨菌草作为培养料，已知可栽培香菇、灵芝等49种食用菌、药用菌。除了作为菌料外，还可作饲料，同时还是水土保持的优良草种。可用于生物质发电、纤维板、制造燃料乙醇等。

in addition to high temperature, can tolerate short-term drought, but not waterlogging.

## APPLICATION,VALUE

Strong stress resistance, high yield, high crude protein and sugar content, one of the high-yield and high-quality fungus grass. Using giant fungus grass as the culture material, it is known that 49 kinds of edible and medicinal fungi such as Shiitake and Ganoderma can be cultivated. In addition to being used as a fungus, it can also be used as feed, and it is also an excellent grass species for soil and water conservation. It can be used in energy applications such as biomass power generation, fiberboard, and fuel ethanol production.

# 75 / 芦荟

阿福花科 Asphodelaceae　芦荟属 *Aloe*

*Aloe vera*

## 植物形态

多年生常绿草本植物。叶簇生，大而肥厚，呈莲座状或生于茎顶，叶常披针形或叶短宽，边缘有尖齿状刺。花序为伞形、总状、穗状、圆锥形等，色呈红、黄或具赤色斑点；花瓣6片；雌蕊6枚；花被基部多连合成筒状。

## 生态分布

原产于非洲热带干旱地区，分布几乎遍及世界各地。在印度和马来西亚一带、非洲大陆和热带地区都有野生分布。在中国福建、台湾、广东、广西、四川、云南等地有栽培，也有野生状态

## PLANT MORPHOLOGY

Perennial evergreen herb. Leaves clustered, large and plump, seated or at the top of the stem, often lanceolate or short and wide, with sharp toothed spines at the edge. Inflorescence umbelliform, racemose, spike shaped and conical, with red, yellow or red spots; petals 6; pistils 6; perianth base connected into tube.

## ECOLOGICAL DISTRIBUTION

Native to tropical arid areas of Africa and distributed almost all over the world. Widely distributed in India and Malaysia, the African continent and the tropics. It is cultivated and wild in Fujian, Taiwan, Guangdong, Guangxi, Sichuan, Yunnan and other places in China.

的存在。

## 生长习性

　　喜光，耐半阴，忌阳光直射和过度荫蔽。适生温度20~30℃，夜间最佳温度为14~17℃。低于10℃基本停止生长，低于0℃芦荟叶肉受冻全部萎蔫死亡。有较强的抗旱能力，离土的芦荟能干放数月不死。芦荟生长期需要充足的水分，但不耐涝。

　　一般采用幼苗分株移栽或扦插等进行无性繁殖。无性繁殖速度快，可以稳定保持品种的优良特征。

## 用途、价值

　　具有杀菌、消炎、湿润、美容、强心活血、解毒、抗衰老、防晒的作用。

## GROWTH HABIT

　　It likes sunlight, resistant to semi shade, avoid direct sunlight and excessive shade. The suitable growth environment temperature is 20-30°C, and the best temperature at night is 14-17°C. The growth of aloe basically stopped below 10°C, and the mesophyll of aloe wilted and died when it was frozen below 0°C. With strong drought resistance, aloe can survive for months. Aloe needs sufficient water during its growth period, but it is not resistant to waterlogging.

　　Asexual propagation is generally carried out by seedling transplanting or cutting. The speed of asexual reproduction is fast, which can stably maintain the excellent characteristics of varieties.

## APPLICATION,VALUE

　　It has the functions of sterilization, anti-inflammatory, wetting, beauty, strengthening heart and blood circulation, detoxification, anti-aging and sunscreen.

# 76 / 芙蓉菊

*Crossostephium chinensis*
菊科 Compositae　芙蓉菊属 *Crossostephium*
别名：香菊、玉芙蓉、千年艾

## 植物形态

半灌木。高10~40cm。上部多分枝，全株密被短柔毛。叶聚生枝顶，狭匙形或狭倒披针形，长2~4cm，宽4~5mm。头状花序盘状，排成有叶的总状花序；总苞半球形；花冠管状，顶端2~3裂齿，具腺点；花冠管状，顶端5裂齿，外面密生腺点。瘦果矩圆形，具棱，被腺点；花果期全年。

## 生态分布

分布于中国、巴基斯坦、菲律宾、日本等地。

## PLANT MORPHOLOGY

Subshrub. Plant height 10-40cm. Upper part with multiple branches, and whole plant densely pubescent. Leaves clustered at the top of branches, and leaves narrow spatulate or narrowly oblanceolate, 2-4cm in length and 4-5mm in width Inflorescence head, discoid, arranged into leafy racemes; involucre hemispherical; corolla tubular, top 2-3-lobed teeth, glandular dots; corolla tubular, top 5-lobed teeth, outside densely glandular dots. Achenes oblong, angulate, glandular dots; flowering and fruiting period all year round.

## ECOLOGICAL DISTRIBUTION

Distributed in China, Pakistan, Philippines, Japan, etc.

## 生长习性

喜温，较耐寒，适生温度15~30℃。喜光，喜潮湿的环境，较耐阴。耐旱、耐大风、耐碱。适生于腐殖质深厚、疏松、排水透气性好、保水保肥力强的中性至微酸性砂质土。

## 用途、价值

抗逆性、适应性很强，广泛用于盐碱地改造。株形紧凑，叶片银白，可用于盆栽绿化；用于制作各种树桩盆景。根、叶可入药，可治风湿关节痛、风寒感冒。

## GROWTH HABIT

Favoring temperature and is resistant to cold. Suitable for 15-30°C temperature. Like the environment of sunshine and humidity, more resistant to shade. It is resistant to drought, strong wind and saline alkali. It is suitable for neutral to slightly acid sandy soil with deep humus, loose, good drainage and permeability, and strong water and fertility conservation.

## APPLICATION,VALUE

Strong resistance and adaptability, widely used in saline-alkali land transformation. The plant shape is compact, the leaves are silvery white, and it can be used for potted greening; used to make all kinds of stump bonsai. Roots and leaves can be used as medicine, which has the effects of rheumatism, joint pain and cold.

# 77 / 香蕉

*Musa nana*

芭蕉科 Musaceae　芭蕉属 *Musa*

别名：金蕉、弓蕉

## 植物形态

　　多年生草本。植株丛生，高达5m。假茎浓绿带黑斑，被白粉。叶长圆形，长1.5~2.5m，宽60~90cm，基部圆，两侧对称，上面深绿色，无白粉，下面浅绿色，被白粉；叶柄粗，叶翼显著，张开，边缘褐红或鲜红色。穗状花序下垂；花序轴密被褐色柔毛；苞片外面紫红色，被白粉，内面深红色，具光泽；雄花苞片不脱落，每苞片内有花2列；花乳白色或稍带淡紫色；离生花被片近圆形，全缘；合生花被片中间2侧生小裂片长，约为中央裂片的1/2。果丛有果150~300个；果稍呈弓形弯曲，长

## PLANT MORPHOLOGY

　　Perennial herb. Plants grow in clusters, up to 5m high. Pseudo stem dense green, with black spots and covered in white powder. Leaves oblong, 1.5-2.5m in length, 60-90cm in width, round at the base, symmetrical on both sides, dark green above, without white powder, light green below, covered with white powder; petiole thick, winged-petiole prominent, open, and the edges brownish red or bright red. Spike shaped inflorescence drooping; inflorescence axis densely covered with brown pubescence; outer surface of the bracts purple red, covered with white powder, and inner surface deep red, with luster; male flower bracts not falling off, with 2 rows of flowers per bract. Flowers milky white or slightly purplish; free tepal nearly round, entire; middle two lateral lobules of the connate

10~30cm，径3~4cm，有4~5棱，先端渐窄；果柄短；果皮青绿色，成熟后变黄；果肉松软，黄白色，味甜，香味浓。无种子。

## 生态分布

分布于中国、巴基斯坦、巴西、泰国、牙买加等地。

## 生长习性

喜湿热气候，在土层深厚、土质疏松、排水良好的地里生长旺盛。要求高温多湿，适生温度20~35℃，最低不宜低于15℃。香蕉怕低温、忌霜雪。

## 用途、价值

香蕉果实味香、富含营养，香蕉花可食用，植株高大常绿，叶片大，与其他植物配置，可形成良好的防噪声屏障，具有良好的景观效果。适应性强，耐高温、抗风、耐盐碱，在干旱的沙荒地区栽植，获得经济效益的同时，具有防风固沙、绿化荒滩的作用。

tepal long, about 1/2 of the central lobe. Fruits 150-300 in the fruit cluster; fruit slightly arched and curved, 10-30cm in length and 3-4cm in diameter. With 4-5 edges and the apex gradually narrows; short fruit stalk; fruit skin greenish green, turning yellow when ripe; flesh soft, yellow white, sweet, with a strong aroma. No seeds.

### ECOLOGICAL DISTRIBUTION

Distributed in China, Pakistan, Brazil, Thailand, Jamaica, etc.

### GROWTH HABIT

It enjoys a humid and hot climate, and grows vigorously in areas with deep soil layers, loose soil, and good drainage. High temperature and humidity are required, with a growth temperature of 20-35°C and a minimum of 15°C. Bananas are afraid of low temperatures and avoid frost and snow.

### APPLICATION,VALUE

Banana fruits have a fragrant and nutritious taste, and banana flowers are edible. The plants are tall and evergreen, with large leaves. When planted in combination with other plants, they can form a good noise barrier and have a good landscape effect. Strong adaptability, high temperature resistance, wind resistance, and salt alkali resistance. Planting in arid desert areas not only achieves economic benefits, but also has the function of windbreak, sand fixation, and greening wasteland.

第四章

# 竹藤

PART 4　BAMBOO
AND LIANA

# 01 / 箬竹

**Indocalamus tessellatus**
禾本科 Poaceae　箬竹属 *Indocalamus*
别名：长鞘茶竿竹

## 植物形态

　　灌木状。秆高0.5~2m，直径0.4~0.8m，节间长10~20cm，有白粉，圆筒形，绿色；节下方有红棕色贴秆的毛环。箨耳无；箨舌厚膜质，截形；箨片大小多变化，易落。小枝具2~4叶；叶鞘紧密抱秆；无叶耳；叶舌高0.1~0.5cm，截形，叶片宽披针形或长圆状披针形，长20~46cm，宽4~10.8cm，中脉或生有一条毡毛，叶缘生有细锯齿。圆锥花序，花序主轴和分枝均密被棕色短柔毛；小穗绿色带紫，呈圆柱形；笋期4~5月，花期6~7月。

## PLANT MORPHOLOGY

　　Shrub. Culm height 0.5-2m, 0.4-0.8m in diameter, internode 10-20cm in length, with white powder, cylindric, green; reddish brown hair rings attached to the culm below the nodes, no sheath ear; tongue thick, membranous and truncated; size of the sheath variable and easy to fall. Branchlets with 2-4 leaves; leaf sheaths tightly clasped with culm; without auricles; ligule 0.1-0.5cm in height, truncate, leaf blade broadly lanceolate or oblong lanceolate, 20-46cm in length and 4-10.8cm in width, midrib or with a felt, leaf margin with serrate. Panicle, inflorescence axis and branches densely covered with brown pubescence; spikelets green and purple, cylindrical; sprouting period from April to May, and flowering period from June to July.

## 生态分布

分布于中国、巴基斯坦等地。

## 生长习性

喜温、喜湿，适生于深厚肥沃、疏松透气、排水良好的酸性土壤。

## 用途、价值

可用于公园绿化。秆可用作竹筷、扫帚柄等；叶可用作食品包装物、茶叶、斗笠、船篷衬垫等，可加工制造箬竹酒、饲料、造纸及提取多糖等；笋可作蔬菜或制罐头。

## ECOLOGICAL DISTRIBUTION

Distributed in China, Pakistan, etc.

## GROWTH HABIT

Favoring temperature and humidity and is suitable for deep, fertile, loose, breathable and well drained acid soil.

## APPLICATION, VALUE

It can be used for park greening. The culms can be used as chopsticks, broom handles, etc.; the leaves can be used as food packaging materials, tea leaves, bamboo hats, canopies liner, etc., and can be used to process Indocalamus wine, feed, papermaking and extraction of polysaccharides; bamboo shoots can be used as vegetables or canned food.

# 02 / 凌霄

## 植物形态

攀缘藤本。茎木质，表皮脱落，枯褐色，具气生根。叶对生，奇数羽状复叶，小叶7~9，卵形或卵状披针形，长2~10cm，侧脉6~7对，无毛，有粗齿。花顶生，疏散的短圆锥花序，花序长15~20cm；花萼钟状；花冠内面鲜红色，外面橙黄色；花期5~8月。蒴果，每果含种子数粒。种子扁平，多数有薄翅。

## 生态分布

分布于中国、越南、印度、巴基斯坦等地。

## 生长习性

喜充足阳光的环境，耐半阴，忌暴晒。生长势强，耐寒、耐旱、耐瘠薄、耐盐碱。适生于排水良好、疏松的中性土壤。

## 用途、价值

花色鲜艳，花期长，是庭园中棚架、花门的良好绿化材料。花、根、茎和叶均可入药，具有治跌打损伤的功效。

## PLANT MORPHOLOGY

Climbing liana. Stems woody, epidermis exfoliated, withered brown, with aerial roots. Leaves opposite, odd pinnate compound leaves, leaflets 7-9, ovate or ovate lanceolate, 2-10cm in length, lateral veins 6-7 pairs, glabrous, with coarse teeth. flower terminal, and inflorescence is 15-20cm in length; the calyx bell shaped; inner surface of corolla bright red, and outside part orange yellow; flowering period from May to August. Capsule, containing several seeds per fruit. Seeds flat, mostly with thin wings.

## ECOLOGICAL DISTRIBUTION

Distributed in China, Vietnam, India, Pakistan, etc.

## GROWTH HABIT

Favoring sunny environment. Resistant to half shade, avoid exposure to the sun. It has strong growth ability. It is resistant to cold, drought, barren and saline alkali. It is suitable for neutral soil with good drainage and loose.

## APPLICATION,VALUE

With bright colors and long flowering period, it is a good greening material for scaffolding and flower gates in gardens. Flowers, roots, stems and leaves can all be used as medicine, which has the effect of treating traumatic injury.

# 03 / 桉叶藤

*Cryptostegia grandiflora*
夹竹桃科 Apocynaceae　桉叶藤属 *Cryptostegia*
别名：橡胶紫茉莉

## 植物形态

木质藤本。具白色乳液。叶对生，革质，长卵形或长椭圆形。花数朵排成顶生的聚伞花序；萼片披针形；花冠漏斗状，裂片5；副花冠的鳞片锥尖。蓇葖果。种子扁平，银白色。

## 生态分布

原产马达加斯加西南部。分布于中国、巴基斯坦等地。

## 生长习性

喜高温湿润、阳光充裕的环境，适生温度20~32℃，耐热不耐寒。

## PLANT MORPHOLOGY

Woody liana. With white emulsion. Leaves opposite, leathery, long ovate or long elliptic. Flowers several, arranged into terminal cymes; sepals lanceolate; corolla funnel-shaped, lobes 5; corolla scales conical. Follicles. Seeds flat and silvery white.

## ECOLOGICAL DISTRIBUTION

Native to southwest Madagascar. Distributed in China, Pakistan, etc.

## GROWTH HABIT

Favoring high temperature, humid, sunny environment. The suitable temperature for growth is 20-32°C, which is heat resistant but not cold resistant.

## 用途、价值

具有观赏价值；可改变沙地环境，具有美化海岸的作用，是防风固沙的先锋植物。植株分泌的乳汁可提取橡胶。

## APPLICATION,VALUE

It has ornamental value. It can change the sandy environment and beautify the coast. It is a pioneer plant for windbreak and sand fixation. The milk secreted by the plant can be used to extract rubber.

# 04 / 金香藤

夹竹桃科 Apocynaceae　金香藤属 *Pentalinon*

*Pentalinon luteum*

别名：蛇尾蔓

## 植物形态

常绿藤本。茎缠绕，具白色乳汁。叶对生，椭圆形，全缘，革质。花腋生，花冠漏斗形，金黄色；春至秋季开花，花期长达3~4个月。

## 生态分布

分布于中国、巴基斯坦、美国和西印度群岛等地。

## 生长习性

喜高温，适生温度20~30℃；栽培处需排水良好；适生于肥沃的腐殖质土或砂质壤土。

## 用途、价值

可盆栽，用于房屋或天台装饰；也适合庭院的小型花架、栅栏、窗前绿化。

## PLANT MORPHOLOGY

Evergreen liana. Stems twining, with white milk. Leaves opposite, elliptic, entire, leathery. Flowers axillary, funnel-shaped, golden yellow; flowering period spring to autumn, and flowering period lasting for 3-4 months.

## ECOLOGICAL DISTRIBUTION

Distributed in China, Pakistan, the United States and the West Indies, etc.

## GROWTH HABIT

Favoring high temperature, the suitable temperature for growth is 20-30°C; the drainage of cultivation site should be good; it is suitable for fertile humus soil or sandy loam.

## APPLICATION,VALUE

It can be potted for house or roof decoration, and also suitable for small flower trellis, fences and window greening in the courtyard.

# 05 / 清香藤

*Jasminum lanceolaria*
木樨科 Oleaceae  素馨属 *Jasminum*
别名：光清香藤、北清香藤、破骨风、破藤风

## 植物形态

攀缘灌木。叶对生或近对生，三出复叶，绿色，光亮；小叶椭圆形、披针形或卵圆形，长3.5~16cm，宽1~9cm。复聚伞花序，常排列呈圆锥状，花朵密集；花萼筒状；花冠白色，高脚碟状；花柱异长；花期4~10月。果球形或椭圆形，长0.6~1.8cm，直径0.5~1.5cm，黑色，干时呈橘黄色；果期6月至翌年3月。

## 生态分布

分布于中国、巴基斯坦、印度、缅甸、越南等地。

## PLANT MORPHOLOGY

Climbing shrub. Leaves opposite or nearly opposite, ternately compound leaf, green, bright; leaflets elliptic, lanceolate or ovoid, 3.5-16cm in length and 1-9cm in width. Compound cymes, often arranged in panicle shape, with dense flowers; calyx tubular; corolla white, high foot disc-shaped; style hetero long; flowering period from April to October. Fruit spherical or elliptic, 0.6-1.8cm in length, 0.5-1.5cm in diameter, black, and orange when dry; the fruiting period from June to March of the following year.

## ECOLOGICAL DISTRIBUTION

Distributed in China, Pakistan, India, Myanmar, Vietnam, etc.

## 生长习性

生长于山坡、灌丛、山谷密林中。

## 用途、价值

根和茎可入药，能祛风湿、活血止痛。

It grows in hillside, shrub, valley and dense forest.

## APPLICATION,VALUE

Root and stem can be used as medicine, can dispel rheumatism, activate blood circulation and relieve pain.

# 06 / 非洲凌霄

紫葳科 Bignoniaceae

*Podranea ricasoliana*
非洲凌霄属 *Podranea*

## 植物形态

　　常绿半蔓性灌木。高1~2m。叶对生，奇数羽状复叶，叶柄具凹沟；小叶9~13枚，长卵形，长4~6cm，叶缘具锯齿，叶柄基部紫黑色。圆锥花序，顶生，花冠漏斗状钟形，先端5裂，粉红或紫红色，有时带有紫红色脉纹；花期秋至翌年春季。

## 生态分布

　　原产于非洲南部。分布于中国、巴基斯坦等地。

## PLANT MORPHOLOGY

　　Evergreen semi-tendril shrub. Plant height 1-2m. Leaves opposite, odd pinnate compound leaves, petiole with concave groove; 9-13 leaflets, long ovate, 4-6cm in length, leaf margin serrate, petiole base purple black. Panicle, terminal, corolla funnel-shaped bell shaped, apex 5-lobed, pink or purplish red, sometimes with purplish red veins; flowering period from autumn to following spring.

## ECOLOGICAL DISTRIBUTION

　　Native to southern Africa. Distributed in China, Pakistan, etc.

## GROWTH HABIT

　　Favoring warm and light, not drought resistant, but heat and frost resistance. Suitable for

## 生长习性

　　喜温，喜光，不耐旱，耐高温，耐霜冻。适生温度18~28℃。适生于排水良好的壤土或砂壤土中。

## 用途、价值

　　枝条柔软，叶片翠绿而密集，在开阔草坪中大面积连片种植，具有较好的景观效果。

temperature is 18-28°C. Favored growing in loam or sandy loam with good drainage.

## APPLICATION,VALUE

　　The branches are soft and the leaves are green and dense. Planted in a large area in the open lawn with good landscape effect.

# 07 / 硬骨凌霄

紫葳科 Bignoniaceae
*Tecoma capensis*
黄钟花属 *Tecoma*
别名：四季凌霄

## 植物形态

常绿、披散藤本。高1~2m。枝绿褐色，常有小瘤状凸起。叶对生，奇数羽状复叶；总叶柄长3~6cm；小叶多为7枚，卵形至阔椭圆形，长1~3cm，边缘有不规则的锯齿。总状花序，顶生；花萼钟状，5裂；花冠漏斗状，橙红色至鲜红色，有深红色的纵纹，长约40mm，上唇凹入；雄蕊突出；花期春季。蒴果线形，长20~50mm，多不结实。

## 生态分布

原产南美洲。分布于中国、巴基斯

## PLANT MORPHOLOGY

Evergreen, scattered liana. Plant height 1-2m; Branches greenish brown and often tuberculate. Leaves opposite, odd pinnate compound leaves; total petiole 3-6cm in length; leaflets mostly 7, ovate to broad elliptic, 1-3cm in length, with irregular serrated edges. Inflorescence racemose, terminal; calyx campanulate, 5-lobed; corolla funnel-shaped, orange red to bright red, with crimson longitudinal lines, about 40mm in length, with concave upper lip; stamens protruding; flowering in spring. Capsule linear, 20-50mm in length, mostly fruitless.

## ECOLOGICAL DISTRIBUTION

Native to South America. Distributed in China, Pakistan, etc.

坦等地。

## 生长习性

喜温、喜湿、喜光；不耐寒，不耐阴；适生于排水良好的砂壤土。

## 用途、价值

用于庭院绿化，可盆栽观赏装饰。叶片繁茂，花期长，适宜用来美化假山、墙垣。茎、叶、花可药用，主治咳嗽、咽喉肿痛、骨折等症。

## GROWTH HABIT

Favoring warm and humid and sunny. Not cold and shade resistance. Favored growing in sandy loam with good drainage.

## APPLICATION,VALUE

Used for garden greening and potted ornamental decoration. It has luxuriant leaves and long flowering period, which is suitable for beautifying rockeries and walls. Stems, leaves, flowers can be used for medicine, mainly for cough, sore throat, fracture and other diseases.

# 参考文献
## REFERENCES

海鹰, 阿布力米提·阿布都卡迪尔, 曾雅娟, 等, 2010. 中国 – 巴基斯坦喀喇昆仑公路沿线植物区系 [J]. 干旱地区研究, 27(4): 545–549.

HUSSAIN A ,2020. 瓜达尔港与区域经济发展 [D]. 长春 : 吉林大学.

刘梦军, 汪民, 2009. 中国枣种质资源 [M]. 北京 : 中国林业出版社.

刘三才, 1992. 巴基斯坦作物科学研究进展 [J]. 世界农业, (2): 23–24.

卢奇, 王继和, 褚建民, 2012. 中国荒漠植物图鉴 [M]. 北京 : 中国林业出版社.

缪勉之, 张仲卿, 方荫才, 等, 1982. 湖南主要经济树种 [M]. 长沙 : 湖南科学技术出版社.

MUSHARAF K, SHAHANA M, 2014. 巴基斯坦 Sheikh Maltoon 区中心植物资源的民族植物学研究 [J]. 植物药与药理学杂志, 3(1):1–8.

韦巧芳, 2017. 瓜达尔港在 "一带一路" 战略中的地位与作用 [D]. 太原 : 山西师范大学.

袁昌齐, 肖正春, 2013. 世界植物药 [M]. 南京 : 东南大学出版社.

张启帆, 2017. 中巴经济走廊研究 [D]. 乌鲁木齐 : 新疆大学.

张卫明, 袁昌齐, 肖正春, 等, 2017. 一带一路经济植物 [M]. 南京 : 东南大学出版社.

中国科学院中国植物志编辑委员会, 1980. 中国植物志 [M]. 北京 : 科学出版社.

中国农业百科全书编辑部, 1989. 中国农业大百科全书 ( 林业卷上 ) [M]. 北京 : 农业出版社.

ANWAR F, LATIF S, ASHRAF M, et al, 2010. Moringa oleifera: a food plant with multiple medicinal uses.[J]. Phytotherapy Research, 21(1): 17–25.

DANIEL E M, 2009. Natural American Medicinal Plants[M]. Portland, London: Timer Press.

KALIM I, 2016. Gwadar Port: Serving Strategic Interests of Pakistan[J]. South Asian Studies, 31.

KHAN M A, SHAUKAT S, SALEEM M, 1999. Diversity of Molluscan Communities of Gwadar East Bay at Baluchistan Coast[C]. Aquatic Biodiversity of Pakistan: Timber Press.

NASEEM S, AHMED P, BASHIR S E, 2012. Geochemistry of sulphate–bearing water of Akra Kaur Dam, Gwadar, Balochistan, Pakistan and its assessment for drinking and irrigation purposes[J]. Environmental Earth Sciences.

PERVEEN S, KHALIL J, 2015. Gwadar–Kashgar Economic Corridor: Challenges and Imperatives for Pakistan and China[J]. Journal of Political Studies, 22.

REDHA A, AL–MANSOUR N, SULEMAN P, et al, 2012, Drought, salinity and temperature response to photosynthesis in Conocarpus lancifolius[J]. Journal of Food Agriculture &

Environment, 9(2): 361–364.

REHMAN S U, KHALID M, ALI A, et al, 2012. Deterministic and probabilistic seismic hazard analysis for Gwadar City, Pakistan[J]. Arabian Journal of Geosciences.

VAN WYK B E, WINK M, 2004. Medicinal Plants of the World: An Illustrated Scientific Guide to Important Medicinal Plants and Their Uses [M]. Porland:Timber Press.

ZENG F J, SONG C, GUO H F, et al, 2013. Responses of root growth of *Alhagi sparsifolia* Shap. (Fabaceae) to different simulated groundwater depths in the southern fringe of the Taklimakan Desert, China[J]. Journal Arid Land, 5(2): 220–232.

# 中文名索引
## PLANT CHINESE NAME INDEX

# 拉丁名索引
## PLANT SCIENTIFIC NAME INDEX